Historians have long recognized that the rebirth of science in twelfth-century Europe flowed from a search for ancient scientific texts. But this search presupposes knowledge and interest; we seek only what we know to be valuable. The emergence of scholarly interest after centuries of apparent stagnation seems paradoxical.

This book resolves that seeming contradiction by describing four active traditions of early medieval astronomy: one divided the year by observing the Sun; another computed the date of Easter Full Moon; the third determined the time for monastic prayers by watching the course of the stars; and the fourth, the classical tradition of geometrical astronomy, provided a framework for the cosmos.

Most of these astronomies were practical; they sustained the communities in which they flourished and reflected and reinforced the values of those communities. These astronomical traditions motivated the search for ancient learning that led to the Scientific Renaissance of the twelfth century.

D1394421

*Astronomies and Cultures in Early Medieval Europe*

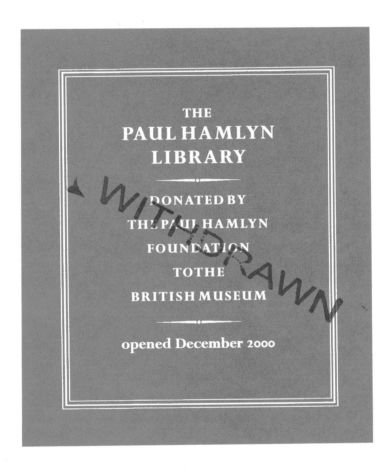

# ASTRONOMIES AND CULTURES IN EARLY MEDIEVAL EUROPE

STEPHEN C. McCLUSKEY
*West Virginia University*

CAMBRIDGE
UNIVERSITY PRESS

PUBLISHED BY THE PRESS SYNDICATE OF THE UNIVERSITY OF CAMBRIDGE
The Pitt Building, Trumpington Street, Cambridge, United Kingdom

CAMBRIDGE UNIVERSITY PRESS
The Edinburgh Building, Cambridge CB2 2RU, UK
40 West 20th Street, New York, NY 10011-4211, USA
10 Stamford Road, Oakleigh, Melbourne 3166, Australia
Ruiz de Alarcón 13, 28014 Madrid, Spain
Dock House, The Waterfront, Cape Town 8001, South Africa

http://www.cambridge.org

First published 1998
First paperback edition 2000
Reprinted 2000

Printed in the United States of America

Typeset in Bembo

*A catalog record for this book is available from the British Library*

*Library of Congress Cataloging in Publication data*
McCluskey, Stephen C. (Stephen Clement), 1940–
Astronomies and cultures in early medieval Europe / Stephen C.
McCluskey.
p.    cm.
Includes bibliographical references and index.
1. Astronomy, Medieval – Europe. 2. Astronomy – History.
3. Calendars – Europe – History. 4. Times – Measurement – History.
I. Title.
QB23.M43      1998
520'.94'0902 – dc21                                    97-6557

ISBN 0 521 58361 6 hardback
ISBN 0 521 77852 2 paperback

# Contents

# Figures

# Abbreviations

| | |
|---|---|
| *AASS* | *Acta Sanctorum* (Paris) |
| *BGPMA* | *Beiträge zur Geschichte der Philosophie des Mittelalters* (Münster i. W.) |
| BL | British Library |
| BN | Bibliothèque Nationale |
| *BOT* | *Bedae opera de temporibus,* (Cambridge, Mass., 1943) |
| *CCCM* | *Corpus Christianorum, Continuatio Medievalis* (Turnholt) |
| *CCM* | *Corpus Consuetudinum Monasticarum* (Sieburg) |
| *CCSL* | *Corpus Christianorum, Series Latina* (Turnholt) |
| *CIL* | *Corpus Inscriptionum Latinarum* (Berlin) |
| *CIMRM* | *Corpus Inscriptionum et Monumentorum Religionis Mithriacae* (The Hague, 1956) |
| CLM | Codex Latinus Monacensis |
| *CSEL* | *Corpus Scriptorum Ecclesiasticorum Latinorum* (Vienna, Prague, and Leipzig) |
| *DSB* | *Dictionary of Scientific Biography* (New York) |
| EETS | Early English Text Society Publications (London) |
| *HAMA* | Otto Neugebauer, *A History of Ancient Mathematical Astronomy* (Berlin, Heidelberg, and New York, 1975) |
| HMML | Hill Monastic Manuscript Library |
| Loeb | Loeb Classical Library (London and Cambridge, Mass.) |
| *MGH* | *Monumenta Germaniae Historica* (Berlin, Hannover, and Leipzig). The parts of this series have been abbreviated as follows: |

| | |
|---|---|
| Auct. Antiq. | *Auctores Antiquissimi* |
| Cap. | *Capitularia* |
| Conc. | *Concilia* |
| Epp. | *Epistolae* |
| Leges. | *Leges nationum Germanicarum* |
| Poet. | *Poetae latini Aevii Carolini* |
| Scr. Rer. Merov. | *Scriptores rerum merovingicarum* |
| SS | *Scriptores* |

| | |
|---|---|
| *PG* | *Patrologia Graeca* (Paris) |
| *PL* | *Patrologia Latina* (Paris) |
| RIG | Recueil des Inscriptions Gauloises (Paris) |

# Preface

The real measure of Christian religious culture on a broad scale must be the degree to which time, space, and ritual observances came to be defined and grasped essentially in terms of the Christian liturgical year.

John van Engen[1]

Early medieval science presents an irritating historical paradox. The conventional view of science in the Hellenistic and Roman periods portrays a declining tradition typified by superficial handbooks and encyclopedias. Conversely, the re-emergence of science in Western Europe began with the active search by medieval scholars for Arab astronomy as early as the tenth century. Yet such active inquiry presupposes knowledge and interest; we seek out only what we already know to be of value. This emergence of scholarly interest after centuries of apparent stagnation defines the paradox.

This book attempts to explain that seeming contradiction by examining early medieval knowledge and practices, attitudes, and institutions reflecting on the heavens. In my discussion of medieval astronomies I have abandoned two common assumptions about astronomy. The first is that of taking the rapid change of modern science as normal and measuring other sciences against some standard of progress. For progress is only half of the picture of science; its goal is not just to expand the realm of the known, but to preserve what is known against error. This is especially true in the natural knowledge of traditional cultures,[2] but even modern scientists, firmly devoted to the advancement of knowledge, spend much of their time passing on what they know to their students.

Thus when we look at the early Middle Ages, we should not consider that the only alternatives were progress or stagnation; rather, they were progress, preservation, or decline. Our question then is not what contributed to progress in astronomy, for episodes of progress were few.[3] Instead, we will ask what forestalled the decline of astronomy and shaped the continuation and renewal of astronomical practice and knowledge from the fourth to the thirteenth centuries.

1. Van Engen, "Christian Middle Ages," p. 543.
2. See, for example, the essays in Wilson, *Rationality*; Horton and Finnegan, *Modes of Thought*; and Hollis and Lukes, *Rationality and Relativism*.
3. Progress was not as rare as suggested by the stereotype of the "Dark Ages" or by one writer who summarized the achievements of medieval astronomers in a chapter consisting of four blank pages. Henry Smith Williams, *The Great Astronomers* (New York: Simon & Schuster, 1930), pp. 99–102. I owe this reference to Owen Gingerich.

We already have the skeleton of an answer, known to anyone who has ever surveyed medieval history. The Christian church, we are told, and particularly the monasteries, preserved classical learning. But can we be more precise: can we determine what kinds of astronomies survived through the early Middle Ages? Can we find what practical or social or ideological functions these astronomies had that made them useful to influential groups in early medieval society and therefore worth preserving? Finally, can we find how these astronomies influenced the search for ancient learning? To what extent can we put meat on these bare bones to see the full countenance of early medieval astronomy?

In seeking to flesh out the skeleton of early medieval astronomy we are forced to challenge the second common assumption: that astronomy can be treated as a single discipline whose practitioners share the same concerns, raise the same questions, and employ the same methods. Rather than limit ourselves to the kind of astronomy that has left its mark on modern practice, I prefer to consider as astronomy any attempt by the members of a community to establish a framework that makes their observations of the heavens intelligible. Since the principal restriction that this definition proposes is that astronomy must be tied, at some point, to observations of the heavens, it leaves room for a wide range of astronomies.

Thus we must begin by distinguishing among medieval astronomical traditions, identifying the problems they dealt with and the astronomical techniques they employed. There are many differences that can help us distinguish different kinds of astronomies, but for the early Middle Ages a few seem most significant. Is the astronomy based purely on observation, or does it use mathematical techniques to predict future observable phenomena? If it does use calculations are they based on geometrical and trigonometric models or on simple arithmetic? Does it trace the continuous motion of bodies, or does it determine the times or places at which individual celestial events occur? If it is concerned with events do these recur once a year or month, and so define a calendar, or every day so as to mark particular times in a day? Which celestial bodies does the astronomy consider: the stars, Sun, Moon, or other planets?[4] Applying these criteria to the early Middle Ages, we can distinguish at least four distinct astronomical traditions.

The first kind of astronomy we encounter is an ancient tradition of dividing the year into eight equal parts using simple observations of the rising and setting Sun. The central concern of this astronomy was to determine ritually and calendrically important dates, but the method was strictly observational. Observers noted when the Sun arrived at a particular point as a way to mark the arrival of

---

4.  I will generally follow ancient tradition and include the Sun and Moon among the seven planets. When discussing details I will often need to distinguish the five starlike planets from the two great luminaries, the Sun and Moon. I will then refer to them as the other planets or, following Ptolemy, the five planets. *Almagest*, 9.1–6, 13.1–4.

a particular day. This astronomy, like other traditional astronomies, had no theoretical framework beyond the simple concept of dividing the year into equal parts.

Easter computus was also concerned with determining calendrically and ritually significant dates, the dates of the Paschal Full Moon and of Easter. These were computed using simple arithmetic and the periods of the Julian year, the lunar month, and the week. Although one result was an observable full Moon, these dates were computed using simple arithmetical techniques in which observations of the Sun and Moon had no direct part. Since the method of computus was fundamentally arithmetical, geometrical considerations of celestial spheres and circles could also be ignored. This does not mean that practitioners of computus never looked at the sky or considered geometrical models of the heavens. Observations and models were often discussed in the computistical literature to illustrate the celestial motions underlying the arithmetical techniques, but they belong more properly to other astronomical traditions.

Monastic timekeeping defined another astronomical tradition, again concerned with determining a ritual time, but here the time was the time of prayer. The techniques employed were observational, watching the course of the stars until they arrived at particular places that marked the time to pray; at first there was little in the way of a theoretical framework to govern these observations. This began to change early in the ninth century as instruments began to supplement simple unaided observations, bringing with them an implicit geometric model of the heavens and an increasingly quantitative measurement of the passage of time.

The final tradition is the one commonly taken as defining ancient and medieval astronomy, the geometrical astronomy of the quadrivium. The central concern of this tradition was the continuous motion of the heavens within a geometrical model of the universe. This tradition gave rise to two related branches. One was a qualitative cosmological model of a geocentric universe with rotating spheres carrying the stars and planets. The other was a predictive geometrical model of circles and epicycles and the mathematical techniques derived from that model to compute the positions of the stars, Sun, Moon, and other planets as functions of time. This tradition was not designed for, and in fact is not ideally suited for, determination of the arrival of particular days or moments that we have seen in the other three traditions.

Any attempt to treat medieval astronomy as a single tradition would pose an unnecessary dilemma. Either we would have to omit major elements of medieval astronomy as outside our concern, or we would find ourselves forcing elements from different traditions into a historically or theoretically inappropriate framework. Nonetheless, while distinguishing among these astronomies we also must remember that no intellectual tradition exists in total isolation; there are instances where these astronomies interact. Computus texts employ horizon observations to illustrate computistical concepts, while school commentaries in the liberal arts draw on data from computistical texts. Despite such interactions, more is to be gained from considering these four traditions separately than from indiscriminately lumping them together.

It is not just these astronomies, their questions and techniques, that concern us. Having identified these different astronomical traditions we find that they flourished in different cultural contexts, made specific contributions to the communities in which they flourished, and reflected and reinforced the values of the communities that supported them.

The traditional solar calendar, in both its prehistoric pagan and its medieval Christian form, was closely tied to local communities. Critical dates in a traditional solar calendar based on local horizon observations were transformed into feasts of important local saints that were sponsored by local elites and represented local, rather than universal, centers of temporal and spiritual power. Assemblies for their feasts on the traditional solar mid-quarter days animated and enriched the local centers of those cults.

While Easter computus also focused on religious rituals, its presentation of a uniform technique for calculating the date of a universal feast reflected the centralizing tendency and desire for uniformity of ritual on the part of the church hierarchy. This standard technique was spread by monastic and cathedral schools and by the Carolingian court, which also disseminated other elements of ritual uniformity.

Monastic timekeeping, like monasticism itself, combined elements of local autonomy with an underlying principle of order regulated by a sacred rule. The divinely ordained order of the stars was observed over local landmarks following local practices, yet all was aimed at following the same orderly round of prayer.

During late antiquity the astronomy of the liberal arts had lost contact, in the Latin West, with its powerful calculating techniques founded on spherical geometry and trigonometry. Since this tradition could no longer compute the precise circumstances of astronomical phenomena, it had lost its strongest connection to the observable sky. But the image it provided of a well-designed cosmos, and the concept that that cosmos is governed by a universal order, had lasting influence in court circles. To rulers the appeal of this image of universal dominion is obvious. That they chose to study this astronomy themselves and display it on ritual regalia and in the art they sponsored, so to be seen as wise rulers, suggests its value in justifying imperial dominion.

Considering these four traditions, the visage of early medieval astronomy becomes richer and more complex. These astronomies not only stimulated the quest for ancient learning from the tenth to the thirteenth centuries, they also provided the framework within which that astronomy was assimilated. Problems arising within the practical astronomies inspired the inquiry into and mastery of this newly recovered learning. These traditions, then, are essential not only for an understanding of early medieval science but also for a complete understanding of the development of science in the later Middle Ages and Renaissance.

In dealing with both popular and learned astronomies, we touch on an increasingly important historiographical issue: the interplay of learned and popular culture. Van Engen has recently surveyed this issue as it applies to the Christianization of western Europe, where the process of conversion is now recognized as a slow adaptation in which Christian and pagan concepts coexisted side-by-side for centuries.[5] In quite different guise this same concern appears in discussions of the folk astronomies of the Americas, where traditional astronomical techniques and their related rituals continued, often with a thin Christian veneer, in native communities.[6]

We find similar survivals in early medieval astronomy. Celtic feasts blend into saints' days; pagan constellations are used to determine the times for monastic prayer; quantitative data from Greek astronomy are employed to compute the date of Easter. The adoption of Christian beliefs and practices under the influence of preexisting pagan rituals, the adoption of Mediterranean geometric astronomy under the influence of traditional techniques, will provide one guiding theme to this discussion.

In tracing this theme I will be mindful of the interactions among different groups in society revealed by studies of the Christianization of Europe and of the close relationships between astronomy, religion, politics, and society revealed by studies of folk astronomies. Just as social historians have drawn insights into the structure of early medieval society from anthropological studies of traditional cultures, so can a historian of science better understand early medieval science by reflecting on archaeoastronomical and ethnographic investigations of astronomies in traditional cultures.

As an overview of a few important aspects of medieval astronomies spanning a millennium, this book cannot claim completeness. Medieval astronomy reminds me of a Mesoamerican jungle: here and there a few mounds indicate the sites of ruins, some of which have drawn explorers, while others remain to be investigated; much however has crumbled into ruin, been looted by scavengers, or lies forever buried. I have tried to find some pattern in the more readily visible monuments, but no doubt future investigators will examine those I have overlooked and find further paths connecting them. In particular, since the practice of astrology only begins to emerge at the end of this period, I have not dug deeply there. Something beyond Thorndike's monumental survey is still needed before we really understand medieval astrology.[7]

---

5. Van Engen, "Christian Middle Ages."
6. Broda, "Mesoamerican Agricultural Calendar."
7. Tester rightly criticized Thorndike and others who saw astrology where none was present through

It is a commonplace to acknowledge the inspiration of one's colleagues and teachers. Although I am taking something of a new approach, this debt is even more apparent. The question this book outlines has been nagging me since my first year as a graduate student, when the late William Stahlman introduced me to folk astronomies. The tools to attempt an answer I owe to David Lindberg and William Courtenay, who introduced me to medieval science and philosophy. My general framework of how astronomies, religions, and cultures interact has been refined by discussions with my friends and colleagues who do archaeoastronomy, especially Anthony Aveni, Johanna Broda, John Carlson, Davíd Carrasco, David Dearborn, and Clive Ruggles. As I placed elements of medieval astronomy in that framework a cautionary comment of the late Ned Zeena, a sun watcher of the tobacco clan at Walpi Pueblo, often came to mind. He recognized how I could see what he was doing as astronomy, but he looked at it as his religion. I have tried to be sensitive to such differences of perception.

One theme of this book is the support of learning through the Middle Ages; the book's appearance demonstrates that such support continues. This book first took form in 1988, during a sabbatical granted me by West Virginia University; years later its publication was assisted by a subvention granted by the university and its Eberly College of Arts and Sciences. Its completion would have been impossible without the assistance of the librarians and curators who granted me access to books and artifacts in their collections, frequently providing me with photographs to illustrate my discussion. I especially thank the tireless interlibrary loan staff of the West Virginia University Library, who cheerfully and efficiently met my requests for esoteric publications, as well as the librarians at St. Vincent Archabbey and College and the University of Pittsburgh, for their neighborly hospitality. Closer to home my son, Tom, brought order to my boxes of photocopies, my wife, Connie, brought clarity to my writing, and young Rose excused my absences. As the book developed, my students and colleagues commented on early drafts; their questions provoked me to define my argument more explicitly. Finally, I must thank the members of the community of medievalists who have edited texts, described artifacts, and looked in detail at aspects of my problem. The bibliography is not just a tool for those who read this book; it catalogs many of my debts. Perhaps this book will be accepted as partial payment.

---

"confusion of *astrologia* [often meaning astronomy] with astrology and the inference to the presence of astrology from descriptions or representations of the zodiac." Tester, *A History of Western Astrology*, pp. vii, 142. Flint's contention that the needs of religious timekeeping led churchmen to preserve astrology involves inferences of that kind. Flint, *Magic in Early Medieval Europe*, pp. 137, 142.

# PART ONE

# *The environment for medieval astronomies*

# CHAPTER ONE

# *Astronomies in cultures*

> With regard to virtuous conduct in practical actions and character, [astron-
> omy], above all things, could make men see clearly; from the constancy,
> order, symmetry and calm which are associated with the divine, it makes its
> followers lovers of this divine beauty, accustoming them and reforming their
> natures, as it were, to a similar spiritual state.
>
> Ptolemy, *Almagest*[1]

The lights in the heavens, the Sun and the Moon, the stars and the other planets, have enticed people to contemplate them from the beginnings of recorded history.[2] In the introduction to his great work of mathematical astronomy, the Hellenistic astronomer Claudius Ptolemy (ca. 100–ca. 175) put his finger on one of the timeless appeals of the heavens. The heavens display a constancy and an order, a symmetry and a calm, that stand as silent challenges to the transience and discord, the irregularity and turbulence, of the world in which we live. He spoke truly when he declared these characteristics to be divine.

Lest I mislead readers who may be unfamiliar with Ptolemy, his work, and its influence, he was not primarily concerned with the ethical symbolism and import of the orderly motions of the unchanging heavens. His *Almagest* was, from beginning to end, a work of mathematical astronomy, providing detailed geometrical models of the harmonious motions of the stars, Sun, Moon, and other planets. Yet the Latin West lost this mathematical astronomy in late antiquity. The only element of Ptolemy's astronomy that had any substantial influence through the early Middle Ages was that general concept, which he shared with philosophers and theologians, of the heavenly spheres as a model of order.[3]

For that reason this book, unlike Ptolemy's, is only peripherally concerned with the theoretical side of astronomy; we will touch on it briefly insofar as we need to know a bit of theory if we are to understand how astronomy was put into practice. Central to my discussion will be the practical aspects of astronomy

---

1. Ptolemy, *Almagest*, 1.1.
2. Apparent tallies of the waxing and waning of the Moon have been found on bones dating as early as the thirtieth millennium B.C.; Alexander Marshak, *The Roots of Civilization* (New York: McGraw-Hill, 1972). But see Francisco d'Erico, "Palaeolithic Lunar Calendars: A Case of Wishful Thinking," *Current Anthropology*, 30(1989):117–118, and Alexander Marshak, "On Wishful Thinking and the Lunar Calendar," *Current Anthropology*, 30(1989):491–500 (with a reply by d'Erico).
3. Taub, *Ptolemy's Universe*, esp. pp. 135–153.

in both the modern and the ancient senses: the uses of the regular motions of the heavens to reckon the passing of times and seasons, and the attempts to incorporate those celestial virtues of stability and order into human lives and societies.

Our concern with these practical uses to which an understanding of the heavens may be put leads to an approach which may disturb some students of mathematical astronomy. Practical astronomy is an art, not unlike the art of the potter, the smith, or the healer. As such, it can exist independently of any formal articulation of a mathematical astronomy or a philosophical cosmology. While modern technology depends upon modern science and cannot exist without it, even now the arts and crafts are only loosely tied to scientific theory. Thus we will more often find ourself discussing astronomical practices concerned with knowing how to determine a specific time, than with astronomical investigations concerned with finding how the heavens move.

### Times and calendars

Studies of a broad range of societies have shown that the passage of the year commonly guides many human activities. The literature is filled with discussions of practical calendars that guide the annual cycle of planting and harvesting crops or of breeding and slaughtering livestock. This concern with observing natural phenomena in order to mark the turnings of the year came before the rise of settled agricultural communities and the development of writing. Even migratory bands of hunters and gatherers knew in detail the seasonal appearance of wild plants and animals as local vegetation bears its fruit and foliage and as birds, fish, and other animals migrate from place to place with the seasons. But, as social beings humans have more complex needs, and they must know the special times of gatherings to hunt, to trade, to negotiate, or to celebrate.[4]

This concern to mark special times does not require an astronomical calendar. The migrations of birds and other animals, the flowering of plants and the appearance of their fruit, all mark the arrival of special times. However, the orderly, cyclical recurrence of astronomical phenomena provides reliable, and widely used, indicators of the passing of the seasons; watching the Sun, Moon, and stars often goes hand in hand with watching other seasonal changes.[5] It is scarcely an exaggeration to say that astronomy, botany, and zoology are all natural human activities. These provide a range of observable regularities that can be used, and in fact have been used, to mark such special times.

I speak consciously of marking times, in the plural, rather than of measuring time, for the essence of a calendar lies in the demarcation of special days and

4. Hudson, "California's First Astronomers," esp. pp. 55–65.
5. For example, Hesiod, *Works and Days*, 383–90, 564–73; Malotki, *Hopi Time*, pp. 395–405; Tedlock, *Time and the Highland Maya*, pp. 185–190.

seasons that return in a recurring cycle, rather than in the measurement of an unending flow of undifferentiated moments of time. There is a temptation to dismiss this attention to special, distinct events, marked by "discontinuous time-indications," as a primitive antecedent of a "genuine system . . . of time-reckoning."[6] This misses the point; the determination of the arrival of the proper times for recurring practical, social, or ritual activities is a continuing human problem. What is central is that there is something, some activity, that sets these days apart from all others and makes them special. If we are to understand how medieval people used astronomy to punctuate the regular flow of the year, we must consider the simple astronomical phenomena they used to mark the return of significant days and to delimit the turning of the seasons.

The Sun is the most potent of heavenly beings, the source of warmth and light, the true fountainhead of life. Its risings and settings delimit the nights and days; its comings and goings mark the turning of the seasons. In diverse cultures and epochs religious thinkers and philosophers, hunters and farmers, have recognized its crucial importance. The annual changes of the Sun's motion separate seasons of growth from seasons of dormancy, seasons of planting from seasons of harvest, seasons when animals are abundant from seasons when they are absent. Making order of this central aspect of human experience – the changing of the seasons – drew the attention of observers to the heavens.

The Sun's annual journey is most clearly divided by its arrival at the limits of its travels – the southern limits which mark the onset of winter in the northern hemisphere and the northern limits which mark the onset of summer. Since the Sun is the source of heat and light, its travels not only provide a sign of the changing seasons but are often seen as causing them as well. Thus its journey is commonly at the heart of an observer's attention.[7]

We commonly note how the days become longer as we move towards summer and again become shorter as we turn towards winter. Usually, the longest day is taken as marking mid-summer and the shortest day, mid-winter. The day-to-day change, however, is quite small. At Mediterranean latitudes, the length of daylight changes by only six hours in the six months from June to December, an average change of only two minutes per day. The change becomes imperceptible at the extremes of the Sun's travels in June and December and so is even less suitable to mark the turning of the seasons. Thus, while the length of day indicates the general progress of the seasons, it cannot mark the seasons precisely without an accurate means of measuring time. It is not surprising then, that while many societies recognized the changing length of day and night and the complications

6. Nilsson, *Primitive Time-Reckoning*, pp. 8–10, 355–358.
7. Some peoples whose calendars were regulated by the stars came to consider the summer stars rather than the Sun as bringers of warm weather. Nilsson, *Primitive Time-Reckoning*, pp. 144–145.

that it raised for timekeeping, there are no examples of societies that measured this change as a way to determine the changes of the seasons.

A second common sign of the changing seasons is the changing height of the Sun at midday. We are used to feeling the noontime Sun beat down on our heads in summer, whereas we cast long shadows across the snow in winter. But it is hard to move from this general observation to precise determination of the end of one season and the beginning of the next.

We can most easily estimate the Sun's changing height in the sky by considering changes in its shadow. We do not need a standard unit of measure; all that is required is a marked stick or a careful pace. The precision with which these day-to-day changes in the shadow can be observed increases when larger objects cast the shadows or when more precise measuring rods are used.

As with the duration of daylight, the noontime height of the Sun changes most rapidly in spring and autumn, and more slowly in June and December, which makes such observations poorly suited for directly marking these turning points in the calendar. However, the changing length of shadows during the day was commonly used to mark the daily passage of time.

The third principal change in the Sun's motions during the year is the changing place of sunrise and sunset. This offers a handy way to find the current position of the Sun. Rather than watching the Sun at noon, we could equally well watch it rising or setting on the horizon. We would soon note that the Sun rises twice a year at any given point between the northern and southern extremes of its annual path along the horizon. Only a short step separates noting that the Sun rises or sets at a particular point on the day of a particular seasonal event from the identification of that point as a sign of that event. Besides such contingent seasonal events as planting and harvesting, the extremes, midpoints, and other regular divisions of the Sun's travel can mark special places on the horizon.

Note that just as a calendar is concerned with special days rather than with the measurement of duration in undifferentiated units of time, so is a solar horizon calendar concerned with special places rather than with the measurement of az-imuth in quantifiable angular units. As we map special times onto special places, these special places can take on names and qualities appropriate to those events. If a place marks the time to plant beans it could become "bean planting mound"; the place of sunset when the blackbirds return could be called "blackbird's wing"; and the place where the Sun rests at the limits of his travels could be the "Sun's house."[8] When the Sun rises at a sacred time, and in most cultures the turnings of the seasons typify such sacred times, the place where he rises is thereby sanc-tified. The constant and regular return of the Sun to this landmark reinforces the sacred nature of the place and of the sacred moment in time.[9] Conversely, since

8. McCluskey, "Historical Archaeoastronomy," pp. 32–40, 47–48; Malotki, *Hopi Time*, pp. 427–441.
9. McCluskey, "Calendars and Symbolism."

these sacred places surround the central place from which the Sun is observed, that center, whether temple or village, also takes on something of the sacred and becomes, in a sense, the center of the world.[10]

Thus far, we have considered following the travels of the Sun against a terrestrial framework: a shadow on the ground or a marker on the horizon. But the Sun's seasonal travels through the heavens can also be mapped using the celestial landmarks provided by the stars in the sky.

People who watch the stars at dawn or in the evening twilight soon come to identify specific bright or conspicuously formed groups of stars. They then notice that different stars are visible at each season of the year. It is by something close to direct observation, to an intuitive process of association, that we connect the bright stars in Orion with winter.[11] The stars are not perceptibly warm, like the Sun. There is no obvious reason to associate them with the seasons; they just seem to be related. The seasonal appearance of these constellations, like the changing height of the noonday Sun, provide general signals for the changing of the seasons.

For more precise indications of the calendar we must connect the stars' regular seasonal cycle of appearances and disappearances more directly with the motions of the Sun. As with the Sun, the most conspicuous seasonal changes of a star's visibility, its first and last appearances, happen when the star is near the horizon. Most striking is the first seasonal appearance of a star or constellation, known technically as its heliacal rising, which occurs shortly before dawn and is thus, in some way, connected with the Sun. Similarly, the last seasonal appearance of a star, the achronical setting, occurs shortly after sunset.

With heliacal rising and achronical setting, we have two well-defined astronomical phenomena that closely connect observations of the stars with the Sun and are also sufficiently sharply defined to be suitable for setting the calendar. Unlike the length of daylight, the height of the Sun, or its position on the horizon, which are best observed when they change rapidly in spring or autumn, these phenomena can provide precise markers throughout the year. The rising or setting of an appropriate star can be observed equally well in spring or summer, in autumn or winter.

But despite the seasonal changes of visibility we see that each star, unlike the Sun, rises at the same point on the horizon throughout the year, moving only slightly in the lifetime of a single person. Here is a constancy, a changelessness, that makes the stars as constant and unchanging a reference frame as the mountains

---

10. In the terms of Mircea Eliade, when the celestial – the Sun – touches the earth at the horizon, such a hierophany reveals a sacred space, a sacred time, and the center of the world. See Eliade, *Sacred and the Profane*, pp. 20–76; *Cosmos and History*, pp. 12–21.

11. The further connection of those stars with the winter hunting season through the role of Orion as hunter suggests the kind of mythologizing by which such associations were preserved through the generations.

on the distant horizon from which they emerge and behind which they set. Furthermore, unlike horizon landmarks, the stars are not confined to a particular locality; their seasonal risings and settings are the same over a wide area.[12]

This relative movement of Sun and stars produces a sharply defined series of days on which stars are first seen. These events provide a sequence of benchmarks to define a solar calendar indirectly by observations of the stars, rather than directly by observation of the Sun.[13] Of course, from another perspective, we could say that the Sun is being observed indirectly against the framework of the stars.

While the Sun and stars provide clear paradigms of regularity, making order of the motions of the Moon and of the five planets requires greater effort. The phases of the Moon recur every twenty-nine or thirty days, but this period has no simple and obvious connection with the annual cycle on which a calendar is based. Nonetheless, since the lunar phases are easily seen, they were widely used to define a secondary unit of time, the lunar month.

Twelve such lunar months total some three hundred fifty-four days, about eleven days less than a solar year. If the series of lunar months are to be synchronized with the solar year a way must be found to decide when to insert a thirteenth month in the calendar. Some societies ignore this problem by using a purely lunar calendar, and others tolerate a communal ambiguity about what month it is (scarcely a decision at all). Some decide when to insert an intercalary month by observing the Sun or stars to determine when a particular solar date fell a month "late" in the lunar calendar, while others define a mathematically regular system of intercalation.[14]

Besides watching the changing shape of the Moon as it goes through its monthly cycle of phases, we also watch the Moon as we do the Sun, noting the changing places where it rises and sets or its changing place among the stars. We will soon notice that the Moon's motion against the backdrop of the stars is limited to the same narrow band of constellations that rise and set near the Sun. Similarly, the Moon rises and sets over the same general range of landmarks as

---

12. Strictly speaking, the appearances of the stars are only constant over a band of constant latitude; they do change as we move north or south.

13. Such observations of the stars do not yield a true solar calendar, as stellar risings and settings recur with the period of the sidereal year (365.2564 days). The changes of the seasons recur with a period of one tropical year (365.2422 days). The difference is perceptible, adding up to a day in less than a century, and will cause problems to a well-regulated calendar. For most agricultural or ritual concerns, significant discrepancies will only arise after a number of centuries.

14. Turton and Ruggles, "Agreeing to Disagree"; Zeilik, "Ethnoastronomy II: Moon Watching"; Neugebauer, *HAMA*, pp. 296–297, 353–357, 584–585.

   Theoretically a thirteenth month should be intercalated every 2.7154 years. Good approximations to this period are found by intercalating 3 months in 8 years (2.6667), 4 months in 11 years (2.75), and 7 months in 19 years (2.7143).

the Sun. In both cases, the Moon follows the same general path as the Sun, although it does wander north and south of the limits of the Sun's path.

At the simplest level, we can view these changing positions as distinctly analogous to those of the Sun, with the Moon returning to its former place along the horizon or among the stars every 27.3216 days.[15] Note that there is a difference of about 2.2 days between this 27 1/3-day sidereal month and the common lunar month (the synodic month) of 29 1/2 days. This means that if the Moon is full when it reaches its farthest north position along the horizon, or a specific place in the constellation Gemini, the next time it reaches that point it will be 2.2 days short of being full, the sixth time it will be 13 days short of being full, which makes it a new Moon, and the thirteenth time, which occurs about a year later, it will be a full Moon once again.[16]

But since the Moon wanders north and south of the Sun's path through the stars, at a more complex level we can concern ourselves with these variations of the Moon's travels. When does the Moon lie precisely on the Sun's path among the stars? When does the Moon reach its northern or southern limits? Where are these limits on the horizon? Where are they in the heavens? Can they be marked or identified in some way? Is there any regular pattern to these variations?

This kind of curiosity readily leads to several interesting, if not immediately practical, results. Eclipses of the Sun and Moon occur only when the Moon lies on the path that the Sun follows through the stars. If we have concerned ourselves with the paths of the Sun and Moon through the stars, we find that the eclipses occur at two points that move slowly along the Sun's path through the stars over 18.61 years.[17] If we have concerned ourselves with changing places of moonrise and moonset on the distant horizon, we find that the northernmost and southernmost points that the Moon reaches every 27 1/3 days are not fixed, but oscillate back and forth around the annual extreme places of sunrise and sunset with the same period of 18.61 years.[18]

The dramatic character of eclipses, and their association with particular points moving through the heavens, may suggest that those points manifest a malign

---

15. We can safely ignore the subtle distinction between the sidereal month, measured against the stars (27.32166 days at A.D. 500), and the tropical month, measured from the vernal equinox (27.32158 days at A.D. 500). The accumulated difference of one day in a millenium would have little impact on practical astronomical observations and calculations.

16. The Moon is full at its northernmost position at what is called the mid-winter full Moon, the full Moon of the winter solstice. The new Moon that appears at this same northern extreme six or seven sidereal months later is the mid-summer new Moon.

17. Similar eclipses in the same constellation must be separated by an integral number of years. The nearest integral value, nineteen years, is also a Metonic cycle which reconciles the lunar and solar calendars. The Babylonians, for one, were clearly aware that similar lunar eclipses, or their avoidance, recurred at the same position among the stars every nineteen years. Cuneiform Text, British Museum 41004, rev. 18–19; as cited in Moesgaard, "Full Moon Serpent," p. 51.

18. Thom, *Megalithic Sites in Britain*, pp. 20–23.

power to obscure the two great luminaries. It would take a somewhat greater level of abstraction to connect the ever-changing positions of extreme moonrise with eclipses and to grant a power to those landmarks. In some cultures, however, these vagaries of the inconstant Moon have been dismissed as irrational.

To summarize the range of possible lunar observations, we can observe the phases of the Moon or its position among the stars or on the horizon to establish a kind of auxiliary calendar. Among these options, it seems unnecessarily complex to relate the Moon with stars or horizon, rather than simply observe the Moon itself. In either case, such simple observations could lead to the further calendric problem of relating the solar year to the lunar month. Finally, long-term observations of the changing position of the Moon can point the way towards an understanding of eclipses.

Observations of the complex motions of the Moon and five planets among the stars or on the horizon, while significant to the development of theoretical astronomy and astrology, had little meaning for those astronomies concerned with the reckoning of time or the keeping of the calendar.[19] Their principal contribution to practical astronomy was in the sense of which Ptolemy spoke. Insofar as mathematical astronomers had demonstrated that the apparently erratic motions of the planets were subject to mathematical laws, this achievement extended the belief in the "constancy, order, symmetry and calm" of the celestial regions, an order that could be used to govern human activities.

19.   For a detailed treatment of early theoretical astronomy, see Neugebauer, *HAMA*.

## CHAPTER TWO

# The heritage of
# astronomical practice

When – Zeus willing – counting from the winter solstice sixty days have
    passed,
then the star Arcturus leaves the sacred stream of Okeanos and first rises
    brilliant at eventide,
then the swallow, shrill voiced daughter of Pandion, flies up into the light
    when the new spring begins;
it is best to prune your vines before her arrival.

Hesiod, Works and Days (ca. VIII c. B.C.)[1]

In his *Works and Days*, Hesiod touches here on many of the methods and functions of early timekeeping: counting the passage of days, watching the Sun, stars,
and birds for signs, and preserving a judicious concern for the gods and one's
crops. Simple practices like these form an important part of the background to
early medieval astronomies, yet the nature of their connection remains uncertain.
In many cases we can establish direct connections between antique traditions and
the Middle Ages; in others we can only point out the prior existence of these
ideas and practices and ask whether, and to what extent, they influenced analogous elements of medieval astronomies.

### Prehistoric solar horizon calendars

The earliest part of the astronomical heritage of the early Middle Ages is not
announced with the eloquence of a Hesiod; it lies concealed in the crude stone
monuments erected in the British Isles during the period from 4000 to 1000 B.C.[2]
Most familiar of these megalithic structures, that is, structures made of large,
roughly hewn, stone blocks, is Stonehenge. Almost everyone has heard, and most
experts accept, that the axis of Stonehenge points towards the place where the

---

1.  Hesiod, *Works and Days*, 564–570.
2.  The seminal studies of British archaeoastronomy are the writings of Alexander Thom; on his
    work and its significance see Ruggles, *Records in Stone*. The best introduction to the issues in
    megalithic astronomy remains Heggie, *Megalithic Science*.

Sun rises at the summer solstice.[3] In addition, a range of further, more contro-
versial, alignments involving other extreme points of sunrise, sunset, moonrise,
and moonset have been proposed (Fig. 1).

Although Stonehenge is the most famous megalithic structure at which such
astronomical alignments have been noted, it is not the only one. But the question
of whether these orientations were deliberate or merely fortuitous called forth
careful consideration of the significance of various forms of evidence. Stonehenge
and other megalithic structures share the ambiguity of most archaeological remains; the possible interpretations that we can assign to them are limited only by
our imagination. Selecting which of these possible interpretations reflect actual
functions served by these structures calls for a greater degree of restraint.

The solar alignments immediately suggested that they marked solar observations
for regulating the calendar. Early investigators noted that these alignments pointed
to the same division of the year at the solstices, the equinoxes, and dates midway
between them that were indicated by folklore and traditional calendars.[4] The
subsequent studies of Alexander Thom went beyond the examination of individual structures to investigate possible alignments at large numbers of sites. Thom's
studies found evidence for the same basic calendric framework and claimed further
indications of subdivisions of the year into as many as thirty-two equal parts[5] and
evidence for extremely precise solar and lunar observations.[6]

Critical examinations of this evidence and new, more systematic surveys of
carefully defined groups of sites have led to rejection of the more extreme of
these claims. The evidence for highly precise "scientific" observations is generally
dismissed as the product of various kinds of unconscious selection of data. What
remains from these more critical analyses is evidence for low-precision calendric
observations of sunrise and sunset to mark the divisions of the year into eight
equal parts and evidence for similar low precision observations of the rising and
setting of the Moon, perhaps associated with religious rituals.[7]

---

3.  Hawkins and White, *Stonehenge Decoded*. Even skeptics now acknowledge the solstitial alignment
    of the avenue extending from Stonehenge; Atkinson, "Archaeoastronomy of Stonehenge."

4.  Lockyer, *Stonehenge*, pp. 21–23, 40–41, 178–190, 203–206, and passim; Somerville, "Prehistoric
    Monuments."

5.  Thom, *Megalithic Sites in Britain*, pp. 102, 107–117. An important characteristic of Thom's analysis
    is that he proposed that the Sun's motions divided the year into equal units of time, rather than
    observing those motions against equal angular divisions of the zodiac. This equal-time model is
    more plausible, as it does not require the production of measuring instruments and development
    of geometrical concepts on which the equal-angle model is based. These two models lead to
    different partitions of the year and different rising points of the Sun on the horizon.

6.  Thom, *Megalithic Lunar Observatories*; Thom and Thom, "Megalithic Lunar Lines."
    These claims for highly precise astronomy suggested to some a centralized society dominated
    by a highly skilled elite of astronomer priests. MacKie, *Science and Society in Prehistoric Britain*. This
    view was at odds with the accepted picture of British prehistory and can no longer be supported
    by the evidence of low precision astronomical alignments.

7.  Heggie, *Megalithic Science*, pp. 165, 181, 222–223; Moir, "Objections to Scientific Astronomy in
    Prehistory."

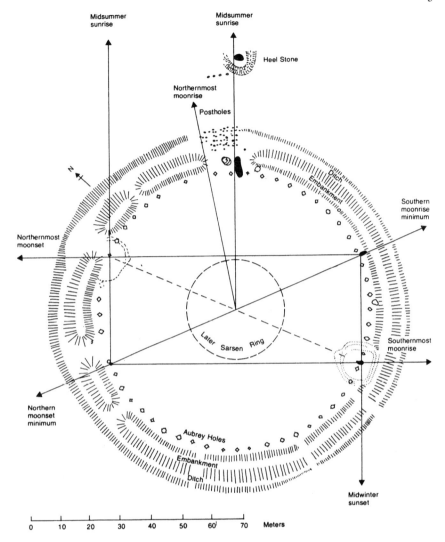

Figure 1. Claimed astronomical alignments at Stonehenge. The alignments are superimposed on a plan of Stonehenge I (ca. 3100 B.C.). Recent excavations have revealed the imprint of a second stone immediately to the left of the Heel Stone, framing summer solstice sunrise between the two. The familiar circle and horseshoe of larger sarsen stones were constructed about 2000 B.C. Reprinted from O. Gingerich, "The Basic Astronomy of Stonehenge," by permission.

More recent studies have revealed the diversity of astronomies practiced at carefully defined and geographically limited groups of sites. Ruggles's study of western Scottish sites showed statistically significant evidence of alignments to the rising and setting Moon at its southern extreme but only slightly more solar calendric alignments than expected and not enough to be statistically significant.[8]

A different pattern emerged from Burl's examination of a cluster of large stone circles in Wales, southwest Scotland, and Ireland. Burl found little evidence for lunar alignments but a greater than expected number of alignments marking sunrise on the dates dividing the solar year into eight equal parts. Archaeological evidence, including trade goods such as stone axes, provided a context for this astronomy by suggesting that these stone circles may have been sites of ceremonial and trading assemblies, for which the solar calendar determined the time.[9]

In summary, these studies of megalithic alignments indicate that the prehistoric peoples of Britain had a tradition of observing the Sun and the Moon. They kept a solar calendar dividing the year into eight equal parts, and performed the kind of lunar observations that could lead logically, but not necessarily, to attempts to reconcile the motions of the Sun and the Moon.

In some respects "megalithic" astronomy does not fit the simple model of an agricultural calendar. Lunar observations are only imperfectly related to the passage of the seasons and seem more related to lunar rituals than to the keeping of a calendar. The solar alignments indicate equal artificial divisions of the year rather than the irregularly spaced, natural times of agricultural activities. The archaeological evidence for ceremonial and trading assemblies at sites where astronomical alignments marked these regular divisions of the year suggests a more complex interaction of astronomy, society, ritual, and trade than that of a simple farmers' calendar.

As we will see, certain elements of the solar calendar established by "megalithic" observers of the heavens survived into the early Middle Ages, despite the cultural changes brought about by successive incursions of Celts, Romans, and Anglo-Saxons. To the extent that we find the same solar calendar playing similar social roles in changing cultural contexts, we can begin to understand how survivals of prehistoric astronomical traditions influenced early medieval knowledge of the heavens.

### Classical horizon systems

The prehistoric Britons were not the only early Europeans to define a system of orientation in terms of the changing position of the Sun along the horizon. Similar

---

8. Ruggles et al., *Megalithic Astronomy*, fig. 12.1, cf. p. 307.
9. Burl limited his study to twelve stone circles having diameters greater than 30 meters. Burl, "Stone Circles of Cumbria," pp. 197–198.

systems were commonly used in Greece and Rome. Antique writers convention-
ally named the winds in terms of the rising and setting points of the Sun at the
solstices and equinoxes, plus north and south, rather than naming directions defined by an equal geometric division of the horizon.[10]

The original system of eight named winds based on observation of the rising
and setting Sun has significant limitations. The places where the Sun rises at the
solstices do not fall at northeast or southeast, but only about a third of the way
from equinoctial sunrise towards due north or south. The large gaps between the
solstitial directions and north or south were often filled in by four additional
directions. This system was embodied in a twelve-sided tower of the winds
erected at Rome, differing from the octagonal tower of the winds in Athens.[11]
Although dominant, these were not the only direction systems in the Roman
tradition; at one point Vitruvius showed how to construct an octagonal wind rose
with only eight equally spaced winds instead of winds based on sunrise and set
at the solstices and equinoxes.[12]

Through late antiquity and into the Middle Ages these directional winds, and
their relationships to the rising and setting Sun, continued to provide architects
with a guide for orienting homes and estates,[13] scholars with a framework for
geographical and cosmic orientation,[14] and artists with ornamental motifs for mo-
saics and other decorative art.

## The stellar calendars of antiquity

If the motions of the Sun provided one way to mark the turnings of the year,
the literary sources found in medieval libraries more commonly described calendars based on the risings and settings of the stars. Although it was not known
in the Middle Ages, the archetype of later stellar calendars is found in the *Works
and Days* of the Greek poet Hesiod, quoted at the head of this chapter. Hesiod
mixes folklore of planting crops, caring for animals, and treating diseases by the
phases of the Moon with advice that the Pleiades define the proper times for
plowing and reaping, whether one lives on the plain, by the sea, or amid a forest.[15]
Here Hesiod alludes to a major advantage of a stellar over a horizon calendar: the
constellations provide celestial benchmarks that, once known, can be used over
a wide area without having to identify new local landmarks everywhere one goes.

Because of this universality, stellar calendars were widely employed and dis-

10.  Pliny, *Hist. nat.*, 2.46.119; Seneca, *Quaest. nat.*, 5.16; Vitruvius, *De architectura*, 1.6.4–5.
11.  Pliny, *Hist. nat.*, 2.36.119; Cetus Faventinus, *De architectonicae*, 2; Noble and Price, "Tower of
     the Winds."
12.  Vitruvius, *De architectura*, 1.6.12–13.
13.  Vitruvius, *De architectura*, 6.4, 6.6.
14.  Destombes, *Mappemondes*, pp. 28, 35–36; Obrist, "Wind Diagrams."
15.  Hesiod, *Works and Days*, 384–390, 564–624, 765–824.

cussed by many Roman writers, who emulated Greek models in this as in other areas. This is not the place to consider all such presentations, but only to sample a few that were read and quoted in the early Middle Ages.[16] The poet Virgil (70–19 B.C.) wrote his *Georgics* as a guide for farmers, and like Hesiod in his *Works and Days*, included weather lore and traditional beliefs about the suitability of phases of the Moon for certain tasks. Yet he went beyond Hesiod in astronomical detail to provide a brief discussion of the nature of the celestial sphere and how the Sun's passage along its slanting path is marked by the twelve constellations of the zodiac.[17] But such knowledge is not enough; Virgil cautioned the farmer to watch the planets and worship the gods. A man is blessed both when he knows the causes of things and when he knows the rustic gods.[18]

The gods are central to another stellar calendar known to the early Middle Ages. The poet Ovid (43 B.C.–ca. A.D. 18) did not direct his *Fasti* to the practical needs of farmers, but addressed ceremonial concerns by marking the stellar signs of civic and religious festivals and describing the gods and events the feasts commemorated, all within the framework of Julius Caesar's reformed calendar. In subject, the *Fasti* is much more an account of Roman mythology and history than it is an astronomical treatise. Yet in structure the *Fasti* is a calendar, and Ovid mentioned in their turn the seasonal risings and settings of the stars, the changing of the seasons, and the solstices and equinoxes. As such, the *Fasti* provided another vehicle for the transmission of these astronomical elements of the Roman calendar to the Middle Ages.

Much broader in scope is the *Natural History* of Pliny the Elder (A.D. 23–79). Like later medieval writers, Pliny did not treat astronomy as a single subject, but discussed various aspects of what we see as a single subject in different sections of his book. Book Two treats cosmology, discussing the nature of the universe, the motion of the stars, Sun, Moon, and other planets, the sphericity of the earth, and related topics. Book Six deals with geographical issues, discussing the size and shape of the earth and the parallels of latitude and noting the changing duration of the longest day and the changing length of the noontime shadow as one goes north of the equator. Book Eighteen considers the basis of the calendar within a general discussion of agricultural practices, in much the same fashion as we have seen in the *Georgics* but surpassing Virgil in detail.[19]

Here Pliny related astronomy to human needs, praising Hesiod, Virgil, and others who had led in this arduous task of "introducing the divine science of the heavens to the ignorance of the rustic."[20] In his discussion of the agricultural calendar, Pliny presented a much more detailed exposition of the calendric prob-

16. Ogilvy, *Books Known to the English*; Laistner, *Thought and Letters*, pp. 218–220.
17. Virgil, *Georgics*, 1.204–258.
18. Virgil, *Georgics*, 1.424–464, 277–286, 335–350; 2.490–494.
19. On Pliny's astronomy and its influence, see Borst, *Plinius und seine Leser*; O. Pedersen, "Astronomical Topics in Pliny," esp. pp. 168–170; and Eastwood, "Plinian Astronomy."
20. Pliny, *Hist. nat.*, 18.56.206.

lem than had his predecessors. He did not limit himself to practical observations that determine the times to plant and harvest, but noted how calendars connect the changing course of the seasons to the motion of the Sun and connect that motion, in turn, to the disappearances and reappearances of the stars.

Besides explaining the general principles underlying the stellar calendar, Pliny worked his way systematically through the year, discussing in turn the activities called for by the rising and setting of each celestial sign. Where Hesiod and Virgil had given single dates for these phenomena, Pliny deliberately presented conflicting data from a wide range of sources:

[T]he morning setting of the Pleiades is given by Hesiod . . . as taking place at the close of the autumnal equinox, whereas Thales puts it on the 25th day after the equinox, Anaximander on the 40th, Euctmeon on the 44th, and Eudoxus on the 48th. We follow the observation of Caesar especially: this will be the formula for Italy: but we also state the views of others, since we are not treating of a single country but of the whole of nature.[21]

Pliny accepted these differences as due to the different regions for which the observations were made. Although he knew in general terms that the times of these phenomena may vary from place to place due to the convexity of the universe and the sphericity of the earth, he avoided the mathematical details required to reconcile or criticize these conflicting dates.[22] This lack of critical judgement among competing sources, characteristic of Pliny and his fellow encyclopedists, coupled with the waning knowledge of the geometrical basis of ancient astronomy, challenged Pliny's medieval readers as much as it enlightened them.

## Geometrical astronomy

Pliny's treatments of astronomy were not limited to calendric materials. He introduced a wide range of less immediately useful astronomical concepts. He distinguished between stars and planets and discussed in vague terms the characteristics and periods of their motions and gave various estimates of their distances from the earth. He described the retrograde motions of the five planets and even suggested a physical explanation of these retrograde motions in terms of rays emanating from the Sun.[23] He explained that eclipses of the Moon are caused by

21. Pliny, *Hist. nat.*, 18.57.212–14.
22. Pliny, *Hist. nat.*, 18.57.210, 216–7; 18.59.220–222. Pliny never goes beyond a rudimentary discussion of the underlying principles to attempt geometric demonstrations of how the spherical shape of the earth influences the risings of stars.
23. Pliny, *Hist. nat.*, 2.6–20. The notion that the Sun's rays cause the retrograde motion of the superior planets is an attempt to explain the easily observed fact that these planets do reverse their motion when they are opposite the Sun.

the earth's shadow and that they recur with a periodic cycle of 223 months.[24] He noted that all seven planets, including the Sun, travel in the band of the zodiac, moving north or south of the center, and that the zodiac is inclined, although he failed to define it or its inclination in reference to any geometric system of co-ordinates. He gave the places in the zodiac of two different values of the planets' *apsides altissimae*, one described as the greatest distance of a planet's circle from the center of the earth, the other as its greatest elevation from its own center.[25] Yet he did not clarify the meaning of these two different concepts, the first of which is related to the eccentricity of the planet's deferent in a geometric model and the second of which is actually the planet's exaltation, the place where it has the greatest astrological influence. Furthermore, he did not relate these to any geometrical model, except to note that the planets appear to be smaller and to move more slowly when farther from the earth. It was such rudimentary under-standing of the terminology and findings of Greek geometrical astronomy, but not of its methods and principles, that Pliny passed on to succeeding generations.

If we are to understand the limits of early medieval astronomy, we must rec-ognize not only what medieval astronomers learned from their predecessors, but what they lost. The foundation stone of ancient astronomy was a geometrical model of the universe in which the stars were viewed as cosmic ornaments on a sphere that rotates around the central earth (Fig. 2). Greek astronomers, like most others, began by placing the observer in a privileged position at the center of the universe.[26] But by choosing the stars rather than a local horizon as their reference, Greek astronomers not only framed a universally applicable astronomy, they also laid the groundwork for mathematical analyses that would be impossible in an astronomy tied to the horizon.

The primary achievement of this spherical model was that it explained the simple observation that most stars rise and set each day at the same points on the eastern and western horizon, while some wheel around in the northern sky with-out setting. For traveling merchants and explorers like the Greeks, this model also explained that as one travels northward on a spherical earth, some stars sink lower in the southern sky until they never rise above the horizon, whereas as one travels south, some stars that cannot be seen from Greece begin to peek above the southern horizon. The stars between the never-setting northern stars and the

24. Pliny, *Hist. nat.*, 2.10.56–7. This so-called Saros cycle of 223 synodic months (approximately 18 years) is more precise than the approximate "cycle" in which eclipses sometimes recur at the same point among the stars after nineteen years (245 synodic months). Eclipses separated by a Saros cycle, however, have moved by some ten degrees among the stars or, what is roughly equivalent, by some ten days in the year.

25. Pliny, *Hist. nat.*, 2.13.63–65. Neugebauer, *HAMA*, pp. 802–803, finds only vague outlines of a planetary theory in Pliny.

26. On the manifold political, philosophical, and mythic significances of the center in Greek thought, see Jean-Pierre Vernant, *The Origins of Greek Thought*, trans. of *Les origines de la pensée grecque* (London: Methuen, 1982), pp. 119–132.

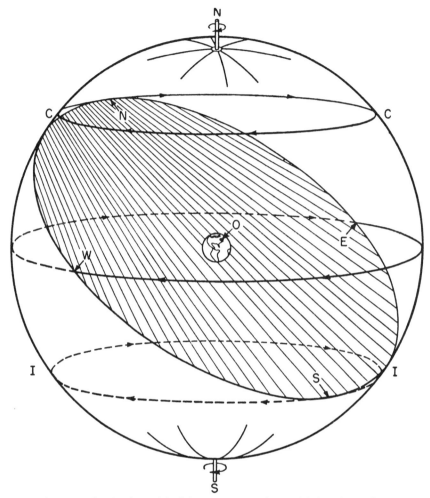

Figure 2. The Greek model of the universe. In this model the sphere of stars rotates daily around the central earth. For an observer at *O*, stars north of the circle *CNC* will never set, stars south of the circle *ISI* will never rise and cannot be seen, and stars between *CNC* and *ISI* will rise at fixed points on the eastern horizon *NES* and set at fixed points on the western horizon *SWN*.

Reprinted by permission of the publishers from *The Copernican Revolution: Planetary Astronomy in the Development of Western Thought* by Thomas S. Kuhn, Cambridge Mass.: Harvard University Press, Copyright © 1957 by the President and Fellows of Harvard College, © 1985 by Thomas S. Kuhn.

never-rising southern stars are above the horizon for varying periods of time, decreasing from north to south.

The spherical model connects our observations of the Sun and the stars by considering that the Sun moves through the stars on an inclined path traced on the celestial sphere. This path, called the ecliptic, goes through those twelve constellations making up the zodiac, which rise and set near the Sun. When the Sun is in the northern portions of the ecliptic it rises in the northeast, passes high overhead at noon, sets in the northwest, and – like the northern stars – shines for more than twelve hours. Conversely, when the Sun is at the southern portions of the ecliptic it rises in the southeast, is low in the noontime sky, sets in the southwest, and shines for fewer than twelve hours.

The Sun traverses the ecliptic once a year, moving in a direction opposite to the daily rising and setting of the starry sphere. Thus the Sun lags behind the stars by about four minutes per day, or looked at another way, the stars move ahead of the Sun by about four minutes per day. This model, then, provides an explanation for the heliacal rising of a star when it emerges from the dawn glare of the Sun and rises before dawn four minutes earlier on subsequent days. Furthermore, since the Sun's travels along the inclined path of the ecliptic take it north and south to produce the changes of the seasons, this model explains why the heliacal rising of a particular star is tied to a particular date in the solar year.

Many of the qualitative explanations of the principal astronomical phenomena provided by the spherical model continued to be taught through late antiquity and the early Middle Ages. But geometry is not merely qualitative. Greek astronomers had used this simple spherical model to compute how long a given star, or the Sun at a given point on the ecliptic, would be above the horizon. More complex geometrical models were analyzed to yield numerical answers to other, more subtle questions about the motion of the Sun, Moon, and other planets. These trigonometric techniques exceeded the mathematical skills of medieval scholars.

## Ptolemaic astronomy

The greatest of the computational achievements of the ancient astronomers was the system of Claudius Ptolemy (ca. 100–ca. 175). His work was so complete that he eclipsed his predecessors, and, as Neugebauer notes, he had no worthy successor. His work continued to provide the basis for a tradition of mathematical astronomy in the Byzantine and Islamic worlds, but in the Latin West he remained little more than a name, often confused with the Ptolemaic rulers of Egypt and styled Ptolemy, king of Alexandria.[27]

Ptolemy's chief astronomical work is titled in Greek the "Megalé Syntaxis"

---

27.   Isidore of Seville, *Etymologiae*, 3.26; Neugebauer, *HAMA*, p. 5.

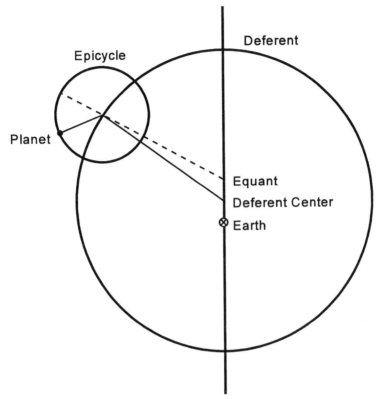

Figure 3. Ptolemy's epicyclic model. This model applies to Venus, Mars, Jupiter, and Saturn; more complex models are required for the Moon and Mercury; simpler epicyclic or eccentric models can be used for the Sun.

but is most commonly known by the Latin transliteration of its Arabic title, the *Almagest*. This reflects the historical fact that for a long time this work was lost to the Latins, who rediscovered it in the Islamic world only in the twelfth century. The *Almagest* is a self-contained treatise on mathematical astronomy, drawing on a discussion of the instruments needed to observe the positions of the stars, and an introduction to trigonometry and spherical astronomy to develop geometrical models for the motion of the stars, Sun, Moon, and other planets based on se-lected astronomical observations extending back some nine hundred years. In these models each planet is carried around on a circle called an epicycle, which in turn is carried around on a larger circle called the deferent (Fig. 3). Ptolemy used these models to prepare tables for a wide range of astronomical computa-tions.[28]

28. For a mathematical and historical commentary, see O. Pedersen, *A Survey of the Almagest*. The

With the *Almagest* one can compute the positions of the stars, Sun, Moon, and other planets for any given time; the time of the solstices and equinoxes and when the Sun enters each sign of the zodiac; the time and characteristics of eclipses of the Sun and Moon; the time of the appearances and disappearances, of the stars; the time of the greatest elongation of Mercury and Venus from the Sun as morning and evening stars; and the time of the appearance, disappearance, and direct and retrograde motions of the five planets. Besides detailing the methods and theoretical bases for such computations, the *Almagest* also includes such practical data as Ptolemy's values for the length of the year, the lunar month, and the seasons.[29]

Ptolemy did not complete his work in mathematical astronomy with the *Almagest*. In his later *Planetary Hypotheses* Ptolemy developed a physical realization of this mathematical model by a system of solid spherical shells, which contained a planet and additional spherical bodies representing its deferent and epicycle (Fig. 4). Since the inner and outer surfaces of these spherical shells were concentric with the earth, Ptolemy could nest them in the order Moon, Mercury, Venus, Sun, Mars, Jupiter, and Saturn to arrive at a cosmos of concentric spherical shells whose outermost dimension was 19,865 times the radius of the earth.[30] His *Handy Tables* consolidated and expanded the *Almagest*'s tables and presented them in convenient form with instructions for their use in astronomical computations, but without the *Almagest*'s geometrical models and demonstrations. Although Ptolemy also wrote lesser works on astronomy, on astrology, and on astronomical instruments, the *Almagest*, the *Planetary Hypotheses*, and the *Handy Tables* define the core of Ptolemaic astronomy.

If we are to understand the later transmission of Ptolemy's astronomy, we should distinguish several levels at which someone could be said to "know" Ptolemaic astronomy. At the simplest level one might know a few astronomical ideas culled, perhaps at second hand, from Ptolemy's works. At an intermediate level one might be able to use some or all of Ptolemy's tables to perform certain astronomical computations, a practical manipulation of numbers that requires little understanding of geometrical astronomy. Conversely, one might know the struc-

---

earliest observation used by Ptolemy is of a lunar eclipse observed at Babylon on the evening of 19 March, 721 B.C.

29.  Following Hipparchus (fl. ca. 162–127 B.C.), Ptolemy computed the tropical year as $365 + \frac{1}{4} - \frac{1}{300}$ days $= 365;14,48^d = 365.246667^d$ and the length of the synodic month as approximately $29;31,50,08,20^d = 29.53059414^d$. He gives the intervals from vernal equinox to summer solstice as 94½ days, summer solstice to autumnal equinox as 92½ days, autumnal equinox to winter solstice as 88⅛ days, and winter solstice to vernal equinox as 90⅛ days. Ptolemy, *Almagest* 3.1, 4.2, 3.4.

30.  *The Arabic Version of Ptolemy's Planetary Hypothesis*, Book I, Part 2.3, ed. and trans. B. R. Goldstein, *Transactions of the American Philosophical Society*, vol. 57, pt. 4 (Philadelphia, 1967), p. 7; Murschel, "Ptolemy's Physical Hypotheses."

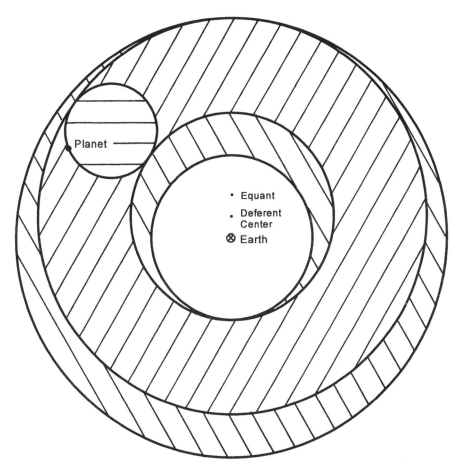

Figure 4. Ptolemy's physical planetary model. Compare this physical realization with the mathematical model of Figure 3. Since the inner and outer
surfaces are concentric, adjacent planetary models can be nested with no intervening empty space.

ture of the Ptolemaic model in which planets are carried around on epicycles
which in turn are carried around on deferents (Figs. 3 and 4) without being able
to perform any astronomical computations. At the highest level one could know
all the details and dimensions of these planetary models, know how to use astronomical instruments to make the observations needed to derive the mathematical
parameters of Ptolemy's models, and know how to use these models to compute
astronomical tables of the sort that Ptolemy provides. Someone at any of these
levels might be called a follower of Ptolemy, but late antique scholars had reached

only the lower levels. They could not pass on the full Ptolemaic tradition to their successors; at most they could pique their successors' curiosity about Ptolemy and his system.

## The decline of observational calendars

If Ptolemy's system was the major achievement of ancient astronomy, the year 46 B.C. marks one of those little-remembered historical changes that undermined the need for practical astronomy. In that year, Julius Caesar introduced a new civil calendar, developed by the astronomer Sosigenes, which throughout the Middle Ages provided the basic framework of the year.[31]

The stability of this calendar must be reckoned as a success of Roman bureaucracy and applied science. Of course, its science is not profound; the notion that the solar year is constant and approximately 365.25 days long is a simple one.[32] The only radical step in the Julian calendar is the adoption of a schematic year in which the months lose all connection to the phases of the Moon.

The application of a schematic calendar to ritual and record keeping would be of little lasting significance if it could not be connected to those aspects of everyday life governed by the changes of the seasons. The Egyptians had used an arbitrary 365-day calendar for such purposes without any great effect on everyday life.[33] But in the Julian calendar the changes of the seasons fell on fixed dates: 25 March, 24 June, 24 September, and 25 December. The calendar could therefore be applied directly to agricultural practice. Pliny, for one, expressed misgivings about the utility of such a mathematical scheme, preferring to compile his stellar calendars for use by farmers.[34]

Nonetheless, the Julian calendar came to be generally accepted so that the farmer no longer needed to watch the changing stars or the movement of the Sun to determine when to plant or to harvest. Telling the time of day still required

---

31. Pliny, *Hist. nat.*, 18.57.211–212. In the ensuing years the Romans reallocated days from one month to the other, while the date on which the year began varied from place to place and time to time. But the basic principle that a cycle of 365.25 days marked a year remained unchanged until the Gregorian reform.

    On the political and social controversies surrounding Caesar's calendar reform, see Bergmann, "römische Kalender." On folk calendars see Eriksson, *Wochentagsgöttter*, pp. 17–37.

32. Sosigenes's choice of 365.25 days has the advantage of falling between the tropical year of 365.2422 days, which measures the changes of the seasons, and the sidereal year of 365.2564 days, which measures the annual appearances and disappearances of the stars.

33. Following the Julian reform, Augustus inaugurated an Alexandrian civil year in which an extra day was inserted every four years in the framework of the Egyptian calendar. The original 365-day Egyptian calendar continued to be used by astronomers such as Ptolemy. Neugebauer, *HAMA*, p. 1064.

34. Pliny, *Hist. nat.*, 18.57.207–217. Pliny's concern is misplaced; his stellar calendar represents the changing seasons less accurately than does the Julian calendar.

astronomical knowledge, but in a society where artificial illumination was limited, most human activity was restricted to the period from dawn to dusk.[35] Practical astronomy was reduced to a simple matter of watching the daily motion of the Sun or its shadow.[36] Knowledge of the stars and of the Moon was no longer essential to the regulation of a civil or farming calendar. The astronomical success in producing an accurate civil calendar had the paradoxical effect of undermining one of the principal practical reasons for further study and observation of the stars.

## Christianity and the Julian calendar

The stable Julian calendar also provided the framework within which the Christian ritual calendar evolved, sometimes in conscious opposition to the pagan Roman ceremonials within that calendar.[37] The Christian calendar, like other ceremonial calendars, was faced with the incommensurability of the annual cycle of the solar calendar, the monthly cycle of the lunar calendar, and the conventional cycle of the seven-day week.

Historically, the first element in the development of Christian ritual was the weekly cycle of worship. The Resurrection, as the central element of Christian belief, was held by the earliest Christians to have sanctified Sunday, on which a gathering for celebration of the Eucharist became the central focus of Christian worship. These early believers saw the fact that Christ rose from the dead on *dies Solis*, the Sun's day, not as an accident, but as a providential sign. For Jerome it was the day on which Christ, the Sun of Justice, *Sol Justitiae* (Mal. 4:2), shone forth.[38] This early aspect of Christian worship in which the Sun was appropriated from pagan ritual and from scriptural texts to provide a spiritual symbol of Christ, flavors much of the early development of the Christian calendar. Thus the weekly assembly on the Sun's day provided a basic element in the later development of the liturgical calendar.[39]

If the Resurrection made all Sundays holy, it especially sanctified that day that marked its anniversary. Thus the earliest Christian communites soon came to

---

35. For a hint of daily life in fourth-century Gaul from cockcrow to bedtime, see Prudentius, *Cathemerinon*, 1–6.
36. This is not to deny that sundials continued to be constructed on mathematical principles in the Roman provinces; see Gibbs, *Greek and Roman Sundials*, nos. 4013, 4014G; but cf. the less appropriately placed 1067, 3104G, 4012, 5020. As successful solutions to an astronomical problem, however, sundials do little to encourage further research.
37. The extensive literature and sources on the encounter of Christianity with solar and lunar cults are examined by Hugo Rahner, *Greek Myths and Christian Mystery*, pp. 89–176. A valuable introduction to the development of the liturgical year is in Vogel, *Introduction aux culte Chrétien*, pp. 263–279.
38. Rahner, *Greek Myths and Christian Mystery*, pp. 107–108; Jerome, *In die dominica paschae*, p. 418.7–19.
39. Vogel, *Introduction aux culte Chrétien*, p. 264.

celebrate Easter, the central feast of the Christian year. Like Passover, with which it was historically linked and which it had replaced for Jewish Christians, Easter was a springtime festival celebrated near the vernal equinox, with the same intimations of the renovation of the world implicit in that season and explicit in contemporary New Year festivals. Easter was variously celebrated on Passover, on the Sunday following or nearest Passover, or even on a fixed date associated with the beginning of spring. We will leave for later further consideration of the complications arising from Easter's ties, through Passover, to the Hebrew luni-solar calendar and Easter's consequent enigmatic variations in the Julian calendar.

The birth of Christ was the next event to be added to the round of Christian ceremonies. Rituals proclaiming that Christ is both divine and human evolved within the contexts of the internal Christian debate over the subtle doctrinal relationship between Christ's divinity and humanity and of the rivalry between Christianity and pagan solar cults.[40] In Alexandria and other places in the Hellenistic world, the birth of Christ was celebrated on 6 January, in opposition to pagan celebrations of the birth of a sun god held on the same date. The growing light on this date, twelve days after the nominal Julian date for the solstice, 25 December, came to symbolize for later interpreters the victory of Christ, the light of the world, and his twelve apostles over the darkness.[41]

A similar conflict with pagan solar rituals arose in the West. In the year 274 the emperor Aurelian had established the cult of *Sol invictus*, the unconquered Sun, as a state cult.[42] The Sun "appears as an infant at the winter solstice, he grows old during the year to appear again as an infant on the shortest day"; hence, the celebration of his birth was at the winter solstice, 25 December.[43] The earliest discussion of a feast on this date that commemorates the birth of Christ is an early fourth-century treatise "On the Solstices and Equinoxes, the Conception and Birth of our Lord Jesus Christ and John the Baptist."[44]

The author set out to demonstrate that the winter solstice is the proper date to celebrate the birth of Christ, taking as his starting point a text from the prophet Zachariah (8:18) that fasts on the fourth, seventh, and tenth months are to be converted to rejoicing and festivals.[45] Adding an implied first month to this list, he interpreted this text as a prophecy of four major festivals at the four turnings of the year: in March, June, September, and December. He then argued that John the Baptist, the precursor of Christ, was conceived at the autumnal equinox (nom-

---

40. An introduction to the literature and sources on the date of Christmas is in Frank, "Frühgeschichte des römischen Weihnachtsfestes."

41. Rahner, *Greek Myths and Christian Mystery*, pp. 137–145. Various sources identify this festival as celebrating either the birth of Dionysius-Aion, son of Kore, or Horus-Harpocrates, son of Isis.

42. Botte, *Origines de la Noël*, pp. 62–63.

43. Macrobius, *Saturnalia*, 1.18.10; Kantorowicz, "Puer Exoriens."

44. De solsticia. On the date see Botte, *Origines de la Noël*, pp. 91–92.

45. De solsticia, 63–66. Zachariah speaks of the fourth, *fifth*, seventh, and tenth months; the author of de solsticia omits the fifth month from his quotation and discussion.

inally 24 September in the Julian calendar) and that the conception of Jesus was six months later at the equinox of spring (nominally 25 March). They were then born, following the natural course of human development, nine months after their conceptions, that is, on the summer and winter solstices (nominally 24 June and 25 December). His discussion draws to an end with a celebration of Christ as the truly unconquered Sun:

[The pagans] also call [25 December] the birthday of the unconquered Sun. Yet who is as unconquered as our Lord who, casting down death, conquered it. They may call this day the birthday of the Sun, but he is the Sun of Justice of whom the prophet Malachi [4:2] spoke: "There shall arise to you who fear his name the Sun of Justice, and there shall be healing under his wings."[46]

At Rome, Christmas had been celebrated on 25 December as early as the year 336.[47] A cycle of four feasts celebrating Christ's birth (Christmas, 25 December) and conception (the Annunciation, 25 March) and St. John's birth (24 June) and conception (24 September) came to endow the solstitial and equinoctial points in the Christian calendar with theological and cosmological symbolism. In his Christmas sermons Augustine drew a Platonic contrast between the pagan worship of the material and visible Sun with the Christian worship of its creator, of Christ, the Sun of Justice.[48]

The relationship of their feast days invited parallels between Christ, the light of lights who "illuminates every soul born into the world" (John 1:9), and the Baptist, the "lamp lit to show you the way" (John 5:35). It soon became a commonplace to compare the two solstices, noting that John and his light declined from his birth at midsummer, while Christ's grew from his birth in the dark of winter. As John said of Jesus, "he must grow greater, while I must diminish" (John 3:30).[49]

By incorporating feasts on the solstices and equinoxes into its ritual life, the Christian church had accepted the astronomical achievement of the Julian calendar

---

46.   De solsticia, 434–9. I have generally followed the translation of Brian Battershaw in Rahner, Greek Myths and Christian Mystery, p. 148.

47.   Botte, Origines de la Noël, pp. 32–33, 62–63; Baumstark, Comparative Liturgy, pp. 153–154. The eastern commemoration of Christ's birth on 6 January came to be celebrated in the West as the feast of the Epiphany. Coebergh, "l'Épiphanie à Rome."

48.   Augustine, Sermones 186, 190.

49.   Augustine, Sermones, 189, 194, 287, 288, (In natali Domini VI, XI, In natali S. Johannis Baptistae I, II); Caesarius of Arles, Sermones, 216; ps.-Bede, De argumentis lunae, PL 90, col. 724. Augustine and Caesarius also expound this passage by noting that John was diminished by being beheaded, while Christ was exalted by being raised on a cross. To modern ears this sounds like an attempt at humor in rather poor taste; one wonders what reverberations it had for Augustine's and Caesarius's congregations.

and had sanctified the calendar and its regularity.[50] Over time this ritual calendar was filled out with an annual cycle of ceremonies. Central to it remained the universal drama of salvation involved in the incarnation, birth, death, and resurrection of Christ. Punctuating this great cycle were lesser ceremonies celebrating events in that drama and in the lives, or more importantly the departure from this life, of a variety of martyrs and other saints. Many of these saints had strong local ties, and their feasts became important community rituals, sometimes continuing older pre-Christian traditions.

Saints' days, as anniversaries of the saints' deaths, became fixed in the Julian calendar. As such, they added a series of significant days against which the various events of everyday life – planting and harvesting, market days and legal assemblies – could be reckoned. If the Caesars had established the Julian calendar throughout their empire, the Christian church would propagate it, transformed by the cycle of Christian worship, beyond the limits of the empire. As Christianity spread, its ritual cycle, now fixed by a rudimentary mathematical achievement of Alexandrian astronomy, would interact with traditional ritual and agricultural calendars based on continued observation and propitiation of the divine Sun and Moon. While in traditional societies the practice of simple astronomical observations had been essential to regulate a community's agricultural practice, public activities, and religious rituals, in medieval Europe calendric observation would become less significant.

---

50.   The continuing significance of this tradition through the tenth century is noted by Ó Carragáin, "Crucifixion as Annunciation," esp. pp. 494–495.

# CHAPTER THREE

# *Astronomy and Christian thought*

Praise him, Sun and Moon; praise him, every star that shines.
Praise him, you highest heavens, you waters beyond the heavens.
Let all these praise the Lord; it was his command that created them.
He has set them there unageing for ever; given them a law which cannot
    be altered.

(Psalm 148:3–6)[1]

The varied encounters of Christianity with ancient astronomical traditions were conditioned by a complex – and somewhat paradoxical – body of Christian teachings about the natural order. Early Christian attitudes towards nature were permeated with the nearness of the miraculous – with God's ability to intervene at will to suspend the natural order. All too often we take this proximity of the miraculous as the sole defining characteristic of Christian perceptions of the natural world. Yet even a cursory reading of Scripture and of early scriptural commentaries reveals a tension between the concept of miracle and the complementary idea of a stable natural order that has been established by divine command.

In the early Middle Ages the Bible came to supplant the works of antique authors as the core of the education of medieval clerics and lay men and women. First, both in sequence and in importance, was the psalter. Children read the psalter and copied verses from it to learn reading and writing; it was used to illustrate a wide range of topics to more advanced students, and many monastic rules required monks to commit all 150 psalms to memory. Knowledge of the psalter, as the touchstone of literacy, produced what Riché has called a psalmodic culture.[2]

It is in the biblical texts, rather than in the philosophical systems that dominated the concerns of the later scholastics, that we will find the key to early medieval attitudes towards nature. For the scholars of the early Middle Ages were not systematic philosophers, teaching in universities; they lived in monasteries, and

---

1.  This and other scriptural citations use the Latin Vulgate text and numbering; English translations generally follow Ronald Knox's modern translation of the Vulgate.
2.  Riché, *Écoles et enseignement*, pp. 222–225; *Education and Culture*, pp. 279–285, 290–293, 463–466.

their lives revolved around the *opus dei*, the daily prayers taken from Scripture, and the *lectio divina*, the regular reading and contemplation of sacred Scripture.

Scriptural texts, by proclaiming an unchanging, divinely established natural order, conditioned medieval perceptions that the heavens reflected that order. In turn, the range of medieval responses to this concept can be found by examining commentaries and other discussions of these texts.

## The scriptural background

A central tenet of the great monotheistic religions is the rejection of polytheism of all sorts. Nature gods and gods of place, and particularly stellar and planetary deities found in earlier mythologies, are supplanted by a single omnipotent God. Natural phenomena are not the result of conflict among competing supernatural agents, but are directed by one ruler whose dominion, in the physical as well as the moral realm, is universal.

This single idea of universal dominion combines two conflicting aspects: one emphasizing the Lord's decrees as the basis of natural law and the other his ability to exercise this dominion to serve his own mysterious ends. In the theological realm emphasis upon a fixed moral decree pushes discussion towards the divine order; the quest for God's purposes in a sometimes meaningless world leads towards the unknowable ways of God. These elaborations of the moral order touch only indirectly on our concern with the order of nature.

More relevant is the order established in the heavens, which appears in biblical texts as a continuing, and growing, theme that emphasizes Yahweh as Creator and Lord of the heavens.[3] In Genesis (1:14–17) the emphasis is merely upon the creation of the great luminaries on the fourth day. To the psalmist it was not just that "the skies proclaim God's glory; the vault of the heaven betrays his craftsmanship" (Ps. 18:2), but that their regular movements reflect the wisdom of God's design (Ps. 103:19–24). This regularity grows in importance to such later writers as the prophet Jeremias (31:35–36) for whom the law that governs the motions of the heavens had become a metaphor for the constancy of the covenant with Israel.

If the heavens are ordered, their intelligibility is more problematic. Since God created the luminaries to be signs of the days and years, we would expect these signs to be intelligible. But to the author of Job, the decrees that "tell the day star when to shine out, the evening star when to rise over the sons of earth" (Job 38:32) are, like all God's judgements, beyond human comprehension. The author of Wisdom however, with his faith that God orders everything by "weight and number and measure" (Wisd. 11:21) prayed for true knowledge of "the disposi-

---

3. This investigation of scripturally conditioned attitudes towards the heavens takes as its starting point the biblical passages dealing with astronomy noted by Allen, *Star Names*, p. 554.

tion of the world, . . . the changes and divisions of the seasons, the course of the year and the position of the stars" (Wisd. 7:17–19).

Yet throughout the Middle Ages, Christian writers would point out cases where God miraculously intervened to alter the established order of nature. The classic examples were the Christmas star (Matt. 2:1–2, 9–10), the darkening of the Sun at Christ's Crucifixion (Matt. 27:45; Mark 15:33; Luke 23:44–45), when Joshua called upon the Sun and Moon to stand still to lengthen the course of battle (Jos. 10:12–14, Sir. 46:5), and when, at the prayer of Isaiah, the shadow on a sundial moved back ten lines (*decem lineis*) as a sign that God would keep his promise to add fifteen years to the life of King Hezekiah (4 Kings 20:1–11; Isa. 38:6–8). The paradox that God had suspended the regular course of the Sun, which he himself had established, to signify that he would keep his promise to extend the normal course of Hezekiah's life was usually resolved to maintain that the created order was unchanging, although human minds might not be able to comprehend it fully.[4]

These contrasting themes of order, intelligibility, and divine dominion appear frequently in the scriptural commentaries, sermons, and other writings of medieval churchmen. This is not to suggest that early medieval commentators intended to present anything like a philosophy of nature; their questions were primarily religious. As Christians they saw God more as redeemer than as lawgiver; as religious thinkers they were more concerned with moral law than with natural law. Yet their responses to scriptural texts outlined a broadly shared attitude towards the order of the heavens. And often, in expounding the various senses of Scripture, they touched on such concepts as the reckoning of time and of the calendar, the structure of the heavens, astrological determinism, and even broader philosophical issues.

### Scriptural commentaries

To the Latin-speaking world, the Bible meant the Vulgate translation of St. Jerome (ca. 327–420). Jerome not only provided the canonical text, he also produced extensive commentaries to clarify the literal meaning of the text.

Discussing the passage where the prophet Jeremias had compared the constancy of God's covenant with Israel to the constancy of his laws, Jerome noted that the order of things, and especially of celestial things, cannot change. The unchangeable laws to which the prophet referred are not the laws given to Moses, but the very constitution and order of nature.[5]

Jerome, not noted for his interest in the natural sciences, nonetheless used a

---

4. Augustine, *De Genesi ad litteram,* 6.16–17; Angelomus of Luxeuil, *In libros Regum,* IV.20. Cf. Hrabanus Maurus, *De clericorum institutione,* 3.25.

5. Jerome, *In Hieremiam,* VI.27.3.

commonplace from the Greek astronomical tradition to compare the globe of the earth on which we dwell to a point, *quasi punctum et globum*, in contrast to those great lights in the heaven whose ordered motions reflect the power of God whom they serve.[6] Although these luminaries are great, they are only creatures, and Jerome condemned Sun worship as an abomination that confuses a creature with the Creator.[7] Similarly, Jerome derided the mythical names of the stars, such as Arcturus and Orion, which reflected the ridiculous lies of the astronomical poets. Nonetheless, he admitted that even he had to use such names for clarity.[8]

Although Jerome recognized an established order in the heavens, when he discussed those times God suspended that order he was more concerned with their spiritual significance than with their physical explanation. He did note, however, that the darkening of the Sun for three hours at the Crucifixion must have been miraculous, for at Passover the Moon was full.[9] He credited the wondrous movement of the shadow on the sundial, and the addition of fifteen years to Hezekiah's life, to Isaiah's calling upon the power of the Lord. It was the divine command, not Isaiah's action, that effected the cure. Yet Jerome did not address the paradox inherent in God's changing an order which he had elsewhere described as changeless.[10]

More influential in forming Christian attitudes towards knowledge of nature was St. Augustine, the learned bishop of the North African city of Hippo Regius from 396 to 430. Augustine's subtle distinctions about astronomy reflect his scholarly background. Discussing a passage (Deut. 4:19) that warns against looking up into heaven and making false gods of Sun, Moon, and stars, Augustine condemned such astrological practices as the casting of horoscopes as vanity; the human soul is not corporeal and subject to any celestial bodies. Nonetheless, he did not reject all study of the heavens, defending as legitimate watching the heavens to determine the course of days and years, "which is commended in Genesis, . . . which use the people of God share with all peoples, but not the [astral] cult, which other peoples have."[11]

Although he sometimes dismissed the importance of natural knowledge for salvation, Augustine interpreted the psalms praising God for his Creation as meaning that one renders the most appropriate praise to God by directing his intellect, made in the image of God, to the works of the Creator. And the works of God are those by which he established the entire created order, an order of rising, an order of setting, an order acting through all time. The study of the heavens, far from being vanity, had become for Augustine a form of worship.[12]

6.  Jerome, *In Esaiam*, XI.40.21/26.
7.  Jerome, *In Hiezechielem*, III.8.15/16.
8.  Jerome, *In Amos*, II.5.7/9.
9.  Jerome, *In Matheum*, IV.27.45.
10. Jerome, *In Esaiam*, XI.38.4/8.
11. Augustine, *De Genesi ad litteram imperfectus*, 3.13; *De Genesi ad litteram*, 2.17, 2.14; *Quaestionum in Heptateuchem*, Lib. 5, Quest. 6.
12. Augustine, *Enarrationes in Psalmos*, CIII, Sermo 4, 2; CXLVIII, 3; CIII, Sermo 3, 25.

Yet while Augustine acknowledged the merit of studying astronomy, he saw limits to its certainty. In his *City of God* he distinguished nature as it is from nature as known, citing both scriptural and pagan accounts of celestial portents that defy astronomical tables (*canones astrologorum*). God's actions are not limited by human experience of the nature of a thing.[13] This distinction took a complex turn in Augustine's discussion of the case of Hezekiah, where he noted that the human intellect can know the hidden natures of things which govern their regular development but cannot know worldly contingencies or things reserved by the unchanging foreknowledge of God. Since God knew from the beginning of time that Isaiah would pray and that he would answer those prayers by adding fifteen years to Hezekiah's life, this action did not disturb the created order.[14]

Knowledge of the heavens took on even greater value in the exposition on the Psalms of the Greek bishop Theodore of Mopsuestia (d. 428), which was soon translated into Latin.[15] We see the Greek confidence in human reason take on Christian garb when Theodore infers from Psalm 18, "The heavens proclaim the glory of God . . . ," that the order of God's creatures proclaims his providence in such a way that all peoples, whatever languages they speak, can confirm the quality of his works through their natural intellect.[16]

Some of the same regard for profane learning appears in the writings of Cassiodorus (ca. 490–583), who sought refuge from the turmoil of the early sixth century in a monastery he founded at Vivarium in southern Italy.[17] Like Augustine, he sometimes introduced philosophical points, discussing the eternity of the heavens, the substance of which the heavens are made, and the concept that the heavens contemplate God with their most pure intellect, inflamed with perpetual love. Aristotle's theory of celestial movers is the ultimate source of the latter idea, but Cassiodorus did not mention Aristotle by name and explicitly rejected the philosophers' belief that the world was eternal. Instead, Cassiodorus took the incorruptibility of the heavens as an example that, as the psalmist said, God's decrees are everlasting. The heavens were created by God, with no confusion or deformity, to carry out his will.[18] This emphasis on divine will meant, for Cassiodorus, that God could suspend his own decrees, as had happened when Joshua

---

13.  Augustine, *De civitate dei*, 21.8.

14.  Augustine, *De Genesi ad litteram*, 6.16–17.

15.  The Latin version of Theodore's commentary exists in two forms, both of which survive in a manuscript formerly at Bobbio and now at Milan. The text was copied by an Irish scribe, Diarmait, who also copied a series of Old-Irish glosses which will be discussed later. Fragments of the translation are extant in two additional codices. L. de Coninck, ed., *Theodori Mopsuesteni Expositionis in Psalmos* (*CCSL* 88 A), pp. vii–xvii.

16.  "Caelorum ergo, dierum, ac noctium non talis est qualis articulatae vocis oratio, – quae frequenter, cum sonaverit, ab alienae linguae hominibus ignoratur, – sed quae omnium linguarum vice fungatur, nullam difficultatem auditus sui apud gentem diversi sermonis experta. . . . Ut possint omnes naturali intellectu factorem de qualitate operis approbare." Theodore of Mopsuestia, *In Psalmos*, XVIII.5$^a$ –5$^b$.

17.  Riché, *Education and Culture*, pp. 158, 161–169.

18.  Cassiodorus, *Expositio Psalmorum*, CXLVIII.1–6, CXXXV.5.

stopped the Sun at Gibeon, when the Christmas star appeared, and when the Sun was eclipsed at the Crucifixion.[19]

Astronomical elements arose even in those scriptural commentaries that favored spiritual rather than literal interpretation of the text. Pope Gregory the Great (ca. 540–604) revealed some familiarity with the heavens when he used Arcturus, which rises in the cold regions to the north, to signify the law, and the Pleiades, which rise in the east, to signify the grace of the New Testament.[20] The more important issues of astrology and astral determinism arose in Gregory's discussion of the Christmas star, where he rejected astrology, maintaining with St. Paul that prophets are given to believers, and signs are only for unbelievers.[21]

A more serious theological problem arose from the heretical Christian astrology of the Priscillianists, who maintained that human souls come from the stars and are dependent upon them; and hence the Christmas star dominated the body of Christ. Gregory argued against such astrological influence, maintaining that the star did not determine the Christ child's fate. Rather, the child fixed the fate of the star. Man is not made because of a star; the stars are made for men.[22]

The rebirth of astronomical learning in Ireland and England during the seventh and eighth centuries, which we will discuss later, was reflected in the commentaries that emerged from that intellectual milieu. The English monk Bede of Jarrow (673–735) usually adopted an allegorical interpretation of Scripture.[23] Yet Bede's extensive study of astronomy led him to justify and employ that discipline when discussing the literal meaning of astronomical allusions in Scripture. His commentary on Genesis shared Augustine's acceptance of the practical use of stars to keep time and to find directions at sea and in the desert. But Bede went beyond Augustine to describe in detail a variety of celestial portents whose use was legitimate.[24]

Bede's commentary on the case of King Hezekiah adopts a strikingly naturalistic tone which he shares with his Irish contemporaries. Bede was more concerned with the nature of the event itself than with the miracle as reflecting God's dominion over nature. He interpreted the text as describing the shadow moving back by an entire day, repeating a day with no nightfall intervening. But, he noted, something like this can be seen "in the island of Thule, which is beyond Britain," where during the night the Sun moves low in the sky from west back to east without setting, although this is never seen in southern regions.[25]

19.  Cassiodorus, *Institutiones*, 2.7.1.
20.  Gregory the Great, *Moralia in Iob*, XXIX.31.73.
21.  I Cor. 14:22; Gregory's text differs from Jerome's Vulgate.
22.  Gregory the Great, *Homiliarum in evangelia*, 10.4. On the astrological doctrines of the Priscillianists, see H. Chadwick, *Priscillian of Avila*, pp. 191–201.
23.  Laistner, "Antiochene Exegesis," p. 29; Smalley, *Bible in the Middle Ages*, p. 36.
24.  Bede, *In principium Genesis*, I.1.14–18.
25.  Bede, *In Regum*, XXV; XXVIII.

The relation between God's dominion and the natural order is discussed in a contemporary gloss on the Psalms by an unknown Northumbrian or Irish author. The glossator made explicit the psalmist's theme that the heavens praise God when they minister to him in their usual order, as well as when they, occasionally, obey him contrary to their nature. He described this order as the "law of nature, [given] to Adam" in contrast to the "law of the letter, [given] to Moses." His gloss also reflects the computists' notion of astronomy as arithmetical, taking God's "numbering all the stars" beyond mere counting to enumerating those numbers by which he governs their course and rising and setting and thereby frees us, presumably from celestial domination.[26]

The Old-Irish glosses on Theodore of Mopsuestia's commentary on the Psalms further attest to the importance of natural learning in insular scholarship. Where Theodore had asserted that all peoples can come to a natural knowledge of God through his creatures, one Irish glossator made it a divine mandate: "it is a law that God should be known through them."[27]

With the growth of monastic schools during the Carolingian renaissance, Anglo-Saxon and Irish scholarship spread to the continent. In his *De clericorum institutione*, Hrabanus Maurus (ca. 780–856), abbot of Fulda and later archbishop of Mainz, advocated the study of astronomy as part of a clerical education. He distinguished superstitious astrology, which he condemned elsewhere, from that natural astrology concerned with finding the times of Easter and other solemnities by considering the movement of the Sun and Moon. Astronomy was especially appropriate for clerics as a contemplative study of the laws of the stars, which cannot move in any other manner than that established by their Creator. In this context, he noted such exceptions to this ordinary course of nature as Joshua commanding the Sun to stand still, the reversal of the Sun by ten divisions (*decem gradibus*) in the time of King Hezekiah, and the eclipse at the Passion of the Lord.[28]

Hrabanus's scriptural commentaries were less original, for like many of his contemporaries, he compiled them from earlier sources to meet the growing pedagogical demand.[29] He quoted without attribution from Bede's commentary on Genesis, repeating verbatim the uses of astronomy and the varied celestial signs that Bede had listed. On Deuteronomy, he appropriated the passage from Augustine that commended watching the heavens to determine the course of days and years, "which use the people of God share with all peoples, but not the [astral] cult, which other peoples have." Astrologers are mentioned in his discussion of the star of Bethlehem, but Hrabanus ignored the star's astronomical import to consider its spiritual symbolism. Similarly, the closest he came to astronomy

---

26. *Glossa in Psalmos*, CXLVIII.1–7, CXLVI.4, introd. pp. 72–73.
27. "Milan Glosses on the Psalms," p. 118.30, 39–40.
28. Hrabanus Maurus, *De clericorum institutione*, III. 25; Hrabanus here excerpts a variant tradition of Cassiodorus, *Institutiones*, 2.7.1. See the condemnation of astrology in Rabanus Maurus, *Poenitentium liber*, 23–24.
29. Smalley, *Bible in the Middle Ages*, pp. 37–38; Laistner, "Some Early Medieval Commentaries."

when discussing the Crucifixion eclipse was a spiritual assertion that Christ's death on the cross occurred at the same hour as Adam's sin.[30]

Pascasius Radbertus (fl. ca. 826–856) drew on a wide range of sources for his lengthy commentary on Matthew's gospel. His discussion of the Christmas star elaborates on Gregory the Great's critique of astrology and Priscillianism, rejecting both the opinion that the star dominated the body of Christ and the specifically Priscillianist view that with the coming of Christ the star became subject to his imperium. Pascasius maintained that the star displayed faithful allegiance, testifying to Christ, the Creator. Thus, he added, this was not a star that had followed a path given it from eternity, but a new star.[31]

Angelomus of Luxeuil (fl. ca. 845–855) composed a similar commentary on the book of Kings, weaving together passages from Bede and Jerome on the nature of Isaiah's miraculous reversal of the sundial as a sign to King Hezekiah. But Angelomus shared Augustine's rejection of any alteration of what God had ordained from the moment of Creation. From the beginning God had intended to grant Hezekiah a longer life than he deserved, although this lay hidden in the counsels of God. God's decrees are unchanging, even if they are not always known to us.[32]

In the introduction to his exposition of Matthew's gospel, Christian of Stavelot (d. post 880) defined three kinds of philosophy to be found in sacred Scripture: natural, moral, and rational. Dividing natural philosophy (or *physica*) into the four arts of the quadrivium: arithmetic, geometry, music, and astronomy, he praised these arts as natural and morally valuable, for "without them . . . we cannot know anything of the good."[33] While he considered astronomy to be valuable, he rejected that divination by the stars practiced by the Magi which, while it may have been permissible before the coming of Christ, cannot now discover any truth.[34]

Christian was primarily concerned with the spiritual, rather than the literal, significance of events. His discussions reflect a cyclical concept of sacred time similar to that in Eliade's anthropological discussions of ritual. He maintained that the Epiphany is important, not only because it was the day on which the Magi, as first of the gentiles, worshipped Jesus but also because *on that same day* John baptized the Lord in the Jordan, Jesus transformed water into wine at Cana, and Jesus fed a great multitude with five loaves. Christian claimed not that these events all happened on the same day in a single year, but that these repeated *Epiphania* occurred on the same day in an annually recurring cycle of sacred time.[35] The

---

30.  Hrabanus Maurus, *In Genesim*, I.5; *Super Deuteronomium*, I.9; *In Matthaeum*, I; *In Matthaeum*, VIII.

31.  Pascasius Radbertus, *Expositio in Matheo*, II. 2.

32.  Angelomus of Luxeuil, *In libros Regum*, IV.20.

33.  "quia istae artes naturales nobis et sine istis mundis non potest consistere, nec nos aliquid boni scire." Christian of Stavelot, *In Matthaeum*, I, *PL* 106, col. 1266.

34.  Christian of Stavelot, *In Matthaeum*, II, *PL* 106, col. 1281.

35.  "non tamen in uno anno, sed per reverentia annorum curricula," Christian of Stavelot, *In Matthaeum*, II, *PL* 106, col. 1286; cf. Eliade, *Sacred and the Profane*, pp. 68–76.

same cyclical sacred time applied to the eclipse at the Passion of Christ, which, as Hrabanus had said earlier, took place *at the same time* that Adam sinned in paradise.[36]

In his commentary on the Psalms, Remigius of Auxerre (d. ca. 908) displays the same interest in the concepts of spherical astronomy and the liberal arts that we see in his commentaries on Boethius and Martianus Capella. In discussing the heavenly creatures that praise God, Remigius distinguished corporeal creatures in the heavens, that is, the seven etherial circles in which the planets and stars are fixed, from the incorporeal creatures, that is, the angels and other celestial spirits.[37]

The broad survey just completed shows that scriptural commentators, rather than opposing the study of the heavens, drew upon contemporary astronomical knowledge in expounding those general biblical themes we first identified. It became customary to accept the practical use of the Sun, Moon, and stars for finding time and directions, since God had established them as signs to govern the day and the night. Some writers went beyond this tolerance to advocate the study of the heavens as a proper, and even obligatory, activity of man as an intelligent creature. Although only a few commentators fully appreciated astronomy as an intellectual discipline among the liberal arts, most recognized its value as a practical activity.

Not all commentators maintained that the order that God had established at the Creation is unchanging, but this idea does recur from time to time and never dies out. Clearly in tension with this are miraculous occurrences, by which God may direct these creatures to act contrary to their natures. But to most medieval writers miracles were more important as signs testifying to the holiness of a saint than as capricious intrusions into the natural order.[38] The commentators' discussions of miraculous interventions thus concentrated on the events' spiritual symbolism more than on their troublesome implications as suspensions of the course of nature.

Some authors, such as Bede, Hrabanus Maurus, and Remigius of Auxerre, who had a special interest in and knowledge of astronomy, took astronomical passages as occasions to display rudiments of this art. Sometimes these occasional discussions presumed on common knowledge of the heavens, shared with their readers, and sometimes they presented less familiar concepts. In either event, they indicate a continued familiarity with some rudiments of astronomy and a continued acceptance of the heavens as an example of natural order.

Since the stars are creatures, we find the commentators formulating variations on the theme that as creatures, they have no dominion over humans. Astrology

---

36.   "Eadem hora qua Adam mortem induxis, eadem ora secundus et verus Adam mortem vicit. Et sicut Dominus tribus horis pependit in cruce, sic Adam tribus horis fuit in paradisio." Christian of Stavelot, *In Matthaeum*, PL 106, col. 1492; cf. Hrabanus Maurus, *In Matthaeum*, VIII.

37.   Remigius, *In Psalmos*, CXLVIII. Smalley sees no further significant commentaries for a century and a quarter after Remigius; Smalley, *Bible in the Middle Ages*, p. 44.

38.   Brown, *Cult of the Saints*, esp. pp. 76–79, 106–127; Ward, *Miracles and the Medieval Mind*.

was generally condemned as a futile superstition, akin to pagan worship of the heavenly bodies. This opposition to astrology was particularly strong among early Christian writers, who were faced with pagan and heterodox Christian forms of Hellenistic astrology.[39] Nonetheless, although various authors defined astronomy and astrology inconsistently, differing on the precise difference between them, they agreed that some kinds of astronomical knowledge and practice were legitimate.

### Astrology and astral religion

One theme usually visible just below the surface of the scriptural discussions of astronomy, and often breaking through to take our attention, was a concern with rival beliefs about the stars. Leading these concerns was that philosophical rival, astrology. Astrology provided educated Romans with a scientific answer to those personal insecurities endemic in the later Republic and Empire, for which others found solace in one of the many Eastern religions sweeping in from the provinces.

Among its Christian opponents, there was a broad consensus that astrology was wrong. They did not, however, establish a single orthodox position but raised, instead, a variety of arguments against astrology. They shared many of these arguments with their pagan contemporaries, insisting, on rational or empirical grounds, that astrology was fallacious. Augustine, like Cicero before him, argued that twins born at the same place and time do not share the same fate.[40] The pagan neo-Platonist Plotinus (ca. 204–270) developed what became a widely held compromise preserving freedom of human action. Although the stars could influence the material human body, they could not influence the immaterial human soul.[41] Christian thinkers added to this the belief that the stars and planets were creatures that followed an order established by God in the beginning, rather than deities or spiritual beings who exercised some kind of independent dominion.

Despite general opposition to an astrological determinism that would deprive humanity of any meaningful freedom, there was greater diversity on the place of astral symbolism, and even elements of astrology itself, in Christian teaching. This appropriation of pagan symbols sometimes developed into a kind of syncretism, especially in Gnostic forms of Christianity.

The prophet's description of the coming Messiah as *Sol Justitiae* (Mal. 4:2), the

---

39.  For general discussions, see Laistner, "Western Church and Astrology"; Bonnaud, "l'Astrologie Latine"; Thorndike, *History of Magic*, vol. 1.
40.  Augustine, *De Genesi ad litteram*, 2.17. Thorndike, *History of Magic*, vol. 1, p. 273, describes the argument from twins in Cicero as a "stock argument," but he gives no earlier example.
41.  Thorndike, *History of Magic*, vol. 1, p. 306.

Sun of Justice, was used widely in Christian art and literature.[42] If the Sun signified Christ, so did the year, of which the psalmist wrote "You crown the year with your bounty" (Ps. 64:12). For some writers this was the year of the public teaching of Christ, extending from his baptism by John in the Jordan to his Passion and Crucifixion. The year-long progress of the Sun symbolized the year-long progress of Christ's teaching.[43]

Since the Sun passes through the twelve signs of the zodiac in the twelve months of the year, the passing of the year was commonly indicated by the number twelve. For some early Christians, just as the Sun signified Christ so were his twelve apostles indicated by the twelve signs, the twelve months, or the twelve hours from sunrise to sunset. As Danielou has pointed out, this theme first arose among Jewish Christians influenced by the Hellenistic writings of Philo of Alexandria, who had developed a similar exegesis comparing the signs of the zodiac with the twelve patriarchs of the Old Testament.[44]

Zeno of Verona (fl. ca. 362–371) sometimes used astral themes in his sermons. He described Christ as our Sun, the true Sun who once set and rose anew and will never set again, crowned with twelve rays, symbolizing the twelve apostles. In a sermon addressed to new converts, being reborn in baptism, he responded to a question about the horoscope of their new birth. Going through the twelve signs of the zodiac, he assigned a spiritual significance to each. For Aries, he substituted a lamb; for Taurus, a little calf; Gemini symbolizes the Old and New Testaments; Cancer the crab becomes an incurable cancer, symbolizing idolatry, lewdness, and greed; and so he continued through the signs to the Scorpion, upon which Christians can tread without harm (cf. Luke 10:19).[45]

If Zeno drew a new spiritual symbolism from astrology, others suggested a more fundamental change. The Gnostic Theodotus (fl. ca. 190) taught that before the coming of Christ, humans had been subject to rival good and evil celestial powers who ruled the course of the stars and planets. Christ at his birth and baptism emancipated believers from these conflicting dominions, destroying the former ordinance of the stars. Against the twelve signs of the zodiac, who governed generation, Theodotus placed the twelve apostles, who were now the agents of that regeneration that liberated Christians from an otherwise implacable destiny. The orthodox censure of Theodotus's concession of even a restricted dominion of the stars over human fate sets a clear limit to acceptable Christian attitudes towards astrology.[46]

---

42. The works of Franz Joseph Dölger thoroughly discuss the themes of Sun and light in early Christian thought. Especially significant is his *Sol Salutis*.

43. Daniélou, "Les Douze Apôtres," p. 18.

44. Daniélou, "Les Douze Apôtres," pp. 19–21.

45. Zeno of Verona, *Tractatus*, 2.12.2; 1.38; Hübner, "Horoskop der Christen"; cf. Clement of Alexandria, *Excerpta ex Theodoto*, 76.2.

46. Clement of Alexandria, *Excerpta ex Theodoto*, 25.2; 69–76; 85.1–3; Daniélou, *Theology of Jewish Christianity*, pp. 223–224; "Les Douze Apôtres," p. 17.

Similar astral beliefs appear among the doctrines attributed to the Spanish bishop Priscillian of Avila, who, in 386, was executed on a charge of sorcery stemming from his unorthodox beliefs and practices.[47] In many respects, Priscillian shared the astral views of more orthodox Christian writers. Like them, he condemned the view that the stars were gods. He criticized even the practice of naming the days and months after the gods as a remnant of this belief, while he accepted the use of the Sun and Moon to reckon the passage of time, since the luminaries were made to serve humanity.[48] Yet some passages suggest that Priscillian also held the syncretistic belief that the power of the stars had been supplanted by the power of the twelve patriarchs.[49]

Considering the astrological teachings of Theodotus and Priscillian, their common deviation from their orthodox contemporaries was in replacing the pagan powers or gods residing in the zodiac with new biblical powers of the patriarchs and apostles. Unlike Zeno, who interpreted the signs of the zodiac symbolically, Theodotus and Priscillian considered the signs as real agents to be replaced by other real agents who, like the signs, acted upon human souls, albeit to liberate them. Despite the variations of Christian attitudes towards astrology, action by the stars on human souls was clearly beyond the limits of Christian orthodoxy.

Coupled with their dismissal of any astrology that saw the stars as effective and independent agents, Christian teachers strongly rejected any astral cult that worshipped the stars as divine. We have already seen one element of this opposition to astral cult in the establishment of a Christian ceremonial calendar in conscious opposition to a cycle of pagan solar festivals. The celebration of Christ's birth confronted the pagan celebration of the birth of *Sol invictus*; the birth of John the Baptist confronted pagan festivals at the summer solstice which Christians were urged to shun.[50] Paganism and Christianity were irreconcilable, in ceremonial cycle as in belief.

Even seemingly innocuous vestiges of astral religion kindled the apprehension of early churchmen. The planetary names of the days of the week: *dies Solis, dies Lunae, dies Martiis, dies Mercurii*, and so on, were an ever-present reminder of the dominion of the planetary deities. Reforms that would replace these pagan names with a neutral numbering of the days were repeatedly advanced, but with little effect. Christian grave inscriptions continued to use the traditional planetary names, and deaths were more frequently recorded on the fortunate days of Jupiter

---

47. H. Chadwick, *Priscillian of Avila*, esp. pp. 132–148, 197–201. A significant contrast with later intolerance is the fact that bishops Martin of Tours and Ambrose of Milan both condemned the bishops who had brought a capital charge against Priscillian. Chadwick, *Priscillian*, pp. 129–130.

48. *Priscillianist Tracts*, I, p. 14.5–19; V, p. 63.25–27; VI, p. 76.19–21; VI, p. 78.3–7.

49. *Priscillianist Tracts*, VI, p. 81.9–13; Paul Orosius, *De errore Priscillianistarum*, p. 153.15–18.

50. Both Augustine, *Sermo 196* (*In natali Domini*), and Caesarius of Arles, *Sermo 33* (*Ante natale Sancti Iohannis Baptistae*), warned their congregations against an unnamed pagan ritual at the summer solstice that involved bathing in springs, in streams, or at the seashore.

and Venus than on the unfortunate days of Mars and Saturn. In Latin and the Romance languages only Sunday, *dies Solis*, took Christian form as *dies Dominici*, the Lord's day. Only in Portugese did Monday through Friday become *segunda-feira* through *sexta-feira*. That the example and recommendation of influential churchmen failed to effect a more widespread change illustrates the deep acceptance of the planetary week.[51]

The most direct threat posed to Christianity by astral religions came from those Oriental cults that shared the monotheistic tendencies of Christianity and Judaism. There was an increasing emphasis on the role of one supreme and transcendant god who governed nature, either directly or through lesser deities. From this perspective the Syrian Baal, the Persian Mithras, and the Hebrew Jehovah are different expressions of a common trend. Considered from the perspective of astral religion, such cults often designated the Sun as the supreme deity, as the first among the subordinate deities, or as a visible symbol of an invisible god.

Most important among these rivals to Christianity, if the wide distribution of temples, monuments, and inscriptions from Asia Minor through Germany to Britain can be taken as an indication, was the cult of Mithras. During the second century the Mithraic cult spread rapidly, extending its early foothold among Roman soldiers and slaves to include merchants and lesser government officials. Under Marcus Aurelius (161–180) the cult was granted civic immunities, and his son and successor, Commodus (180–192), was himself initiated into the Mithraic mysteries.[52]

Conforming to the syncretistic pattern of the times, Mithras was identified with Helios, Apollo, and *Sol invictus*, the unconquered Sun. Cumont notes that the epithet *invictus* was an alien import, generally reserved for the sidereal gods and especially for the Sun. With the growing acceptance of oriental astral cults, the emperors took this potent attribute to themselves. The cult of *Sol invictus* = *Helios* = Mithras thus became an ornament of imperial power.[53] In the year 307 the emperors Diocletian, Galerius, and Licinius dedicated a major temple at Carnuntum, a military headquarters on the Danube frontier, to Mithras as sustainer of their *imperium*.[54]

The powerful Mithraic cult remained a major competitor throughout the conflict between an emerging Christianity and a declining paganism. The religious conflict continued from Constantine's edict of toleration (312) through the brief

---

51.  Caesarius of Arles, *Sermones*, 54.1, 193.4; Martin of Braga, *De correctione rusticorum* 8.17–20, 9.10–13; Bede, *De temporum ratione*, 8.58–63. Eriksson, *Wochentagsgöttter*, pp. 27–33.

52.  My discussion of the cult of Mithras benefits greatly from the work of Cumont, *Mysteries of Mithra*; balanced by Marcel Simon, "Mithra, Rival du Christ?" For recent astral hypotheses of the origins of Mithraism, see Ulansey, *Origins of the Mithraic Mysteries*, and Beck, *Planetary Gods*. On the Mithraic cult's spread in the Western parts of the Empire, see Cumont, *Mysteries*, pp. 39–103.

53.  Cumont, *Mysteries of Mithra*, pp. 97–99; cf. Simon, "Mithra," pp. 467–468.

54.  Cumont, *Mysteries of Mithra*, pp. 88–89, fig. 19; cf. Simon, "Mithra," pp. 469–470.

revival of official paganism under Julian (361–363) and the final effort to breathe vitality into the pagan cults at the end of the century. Throughout this period both the imperial cult of *Sol invictus* and the cult of Mithras continued to have significant, and often overlapping, followings. Reaction to such cults, with their worship of powerful astral divinities, did much to shape early Christian attitudes towards astrology.[55]

Mithras remained a foreign god; his images always portrayed him in Persian attire with trousers, a flowing cape, and a Phrygian cap (Fig. 5). He is frequently depicted in a cave engaged in killing the primordial bull and flanked by two similarly attired torchbearers who, with the central hero, form a "triple Mithras." Cautes, bearing a raised torch, appears in many figures near a bull's head, representing the constellation Taurus, which the Sun enters in late spring; Cautopates, with his torch held downward, is near Scorpio, the constellation which the declining Sun enters in late autumn. Cautes is interpreted variously as the growing Sun of spring and summer, the morning Sun, and a sign of warmth and life; Cautopates represents the declining Sun of autumn and winter, the setting Sun, cold, and death. Many exemplars of this scene add further celestial images: the Sun and Moon, the seven planets, or the signs of the zodiac. In several of them Mithras's billowing cloak is filled with stars to become a representation of the celestial vault and an indication of his power over the cosmos.[56]

Three aspects of the Mithraic cult combined to give Christian apologists an ideal opportunity to attack astral religion and its philosophical cousin, astrology. First was the prevalence of astral imagery, second was the Persian origin of Mithras, and third was the order of Magi who had played a major role in the earlier Persian cult. To Christian ears, Persian Magi who watched the stars had obvious resonances with the Gospel account (Matt. 2:2–12) of the wise men who came from the East bearing gifts for the Christ Child.[57]

This Persian image is reflected and transformed in Christian art from the beginning of the fourth century, as the Magi were added to traditional portrayals of the Nativity of Christ (Fig. 6). The Magi strikingly resemble traditional portrayals of Mithras; they wear the same Persian attire of trousers, flowing cape, and

---

55. Bloch, "Pagan Revival."

56. Beck, *Planetary Gods*, pp. 19–23, 93–94; Cumont, *Mysteries of Mithra*, pp. 130–131; Ulansey, "Mythraic Mysteries." Beck notes that the stellar cloak is found in the Barberini fresco (*Corpus Inscriptionum et Monumentorum Religionis Mithriacae* 390), the Marino fresco, and two sculptures at the Vatican (*CIMRM* 245 and 310); Beck, *Planetary Gods*, p. 83, n. 202.

57. "Ecce magi ab oriente venerunt Hierosolymam dicentes, ubi est qui natus est rex Iudaeorum; vidimus enim stellam eius in oriente et venimus adorare eum." (Matt. 2:2–3).

   The identification of the Magi from the East in Matthew's Gospel with the Persian Magi and the assertion that Christ's appearance broke the domination of the stars appear in Christian writings as early as Ignatius of Antioch (d. ca. 107) and Justin Martyr (ca. 100–ca. 165). Daniélou, *Theology of Jewish Christianity*, pp. 221–224.

   Dieterich, "Weisen aus dem Morgenlande," carries this point further, seeing in the Mithraic astral cult and its connections with the Persian monarchy a germ of the evangelist's story.

Figure 5. Mithras slaying the primordial bull. From the Mithraeum at Heddernheim, Germany. In the center, Mithras and two torchbearers in Persian costume. In the arc above the central figures, the signs of the zodiac. Sammlung Nassauischer Altertümer – Museum Wiesbaden, inv. no. 239. Used by permission.

Phrygian cap. The number of Magi, formerly varying to suit artistic demands, becomes set at three, reminiscent both of the three gifts brought to the Christ Child and of the three figures of the "triple Mithras." Instead of the primordial bull with Mithras's knife at its throat, an ox stands quietly by, illustrating the recognition and worship of the incarnate divinity by dumb nature. Most significant as an anti-Mithraic and anti-astrological portrayal, the Persian Magi them-

Figure 6. Nativity and the Adoration of the Magi. Sarcophagus – Fragment of Cover. Probably from Vatican *Coemitarium*, first third of the fourth century. Magi in Persian attire bearing gifts. The infant Jesus lies in the manger of a tile-roofed stable with ox and ass to the left and a shepherd and Mary to the right. Vatican Museums and Galleries, Lateran Museum, inv. no. 199. Used by permission.

selves recognize the child, reverently offering him gifts symbolic of his kingship, divinity, and humanity.[58]

This same motif is echoed in the writings of the Christian poet Prudentius (b. 348). In his *Hymn for the Epiphany* he placed the Magi in Persia, where through their knowledge of the stars they had recognized the royal standard of "this great ruler who commands the stars, . . . whom light and sky obey." This theme is developed in greater detail in his discourse *On the Divinity of Christ*. Here he placed the Magi symbolically as people of the dawn in the easternmost regions of the Persian realm. Prudentius saw the Christmas star as signifying the overthrow of the wisdom of the astrologers and the power of the stars. Preceded by this new Morning Star, even the Sun sensed its coming eclipse at the Crucifixion of Christ.[59]

Prudentius raised a different critique of astral religion in his attack on the still popular worship of *Sol invictus*, which he addressed to the pagan senator, Symmachus. Prudentius noted that the Sun, day in and day out, is compelled by unerring law to follow the same restricted path, whereas even men can turn from God's decrees. Like the scriptural commentators, he held that the true God is the one who governs all nature; the Sun, constrained to unending servitude, cannot be called a god.[60]

The fourth and fifth centuries saw three related events: the decline of that urban culture which had formed the focus of Roman society; the incursion of the Germanic tribes, which became the dominant political powers in the Western empire; and the emergence of Christianity as the dominant religion. In its expansion Christianity encountered pagan practices that grew as much out of the rustic beliefs of the Germanic and Celtic peoples as they did from the cosmopolitan beliefs of Greece and Rome. The conversion of the countryside only began in earnest in the seventh and eighth centuries; it was not until the thirteenth century that we can say "time, space, and ritual observances came to be defined and grasped essentially in terms of the Christian liturgical year."[61]

Bishops Caesarius of Arles (470–542), Martin of Braga (ca. 515–586), Eligius of Noyon (ca. 588–659), and Pirmin of Reichenau (d. 753) formally warned their flocks against these deeply rooted pagan traditions.[62] They all describe the kind

58. E. Baldwin Smith, *Early Christian Iconography and a School of Ivory Carvers in Provence*, Princeton Monographs in Art and Archaeology, 6 (Princeton: Princeton Univ. Pr., 1918), pp. 13–22, 33–44; Gertrud Schiller, *Iconography of Christian Art* (London: Lund Humphries, 1971), vol. I, pp. 94–101, pl. 57, 144, 146, 147, 149, 151, 246–256; transl. of her *Ikonographie der christlichen Kunst*; Walter Lowrie, *Art in the Early Church* (New York: Pantheon Books, 1947), pp. 81–2.

59. Prudentius, *Cathemerinon* 12.25–37; *Apotheosis* 608–630.

60. Prudentius, *Contra Orationem Symmachi*, 297–353.

61. Van Engen, "Christian Middle Ages," p. 543.

62. Caesarius of Arles, *Sermones*; Martin of Braga, *De correctione rusticorum*; Eligius of Noyon, *Sermo*, pp. 705–708, in Audoenus of Rouen, *Vitae Eligii liber II*; Pirmin of Reichenau, *De singulis libris*

of cult practices that archaeological investigations have associated with the Germanic and Celtic populations of Europe.[63] Masked dances on the Kalends of January (1 Jan.) and rituals at sacred springs, trees, and stones all figure prominently. The mention of sacred stones suggests the kind of rituals associated with lunar and solar observations at stone monuments that we have already noted.

Eligius of Noyon rebuked those who attend to the Moon in the morning before beginning the day's work. This sounds like a practice of noting the Moon's phase or position in the morning, and in it we can see reflections of that lunar calendar keeping common to many folk societies, that is, of the folk traditions connecting the phase of the Moon to the growth of plants and animals or its appearance to the coming weather. Martin of Braga included in his collection of canons a condemnation of observing the course of the Moon or stars according to the traditions of the pagans to determine the time to build a house or plant trees or consort with one's wife.[64]

A more universal criticism was leveled at the practice of raising a great clamor when the Moon was eclipsed, in the belief that this noise would bring back the Moon's light or would protect people from the demons said to come from the Moon when it is darkened. Caesarius and Eligius went beyond mere condemnation to raise the underlying point that the Moon is darkened at fixed time by God's decree, *Deo iubente certis temporibus obscuratur*.[65] This discussion led Eligius to censure those who "call the Sun and the Moon 'lord' and swear oaths by them, because they are God's creatures and serve man's needs at the command of God."[66]

Since the Sun and Moon are not lords, then the practice embodied in the calendar (Fig. 7) of regulating activities by the dominion of the planets over the days of the week must also be rejected.[67] Observing a day of rest on Thursday, Jove's day,[68] or celebrating nuptials on Friday, Venus's day,[69] were noted and condemned. No Christian should be concerned about which day to leave home

*canonicis*. Although the three later writers depended on the work of Caesarius, the fact that they followed Caesarius selectively, sometimes noting new practices, helps us judge when their writings reflect contemporary circumstances.

63. See, for example, Ross, *Pagan Celtic Britain*; Herbert Schutz, *The Prehistory of Germanic Europe* (New Haven: Yale Univ. Pr., 1983), pp. 292–307; *The Romans in Central Europe*, (New Haven: Yale Univ. Pr., 1985), pp. 49–68; Edith Mary Wightman, *Roman Trier and the Treveri* (New York: Praeger, 1971) pp. 208–249.

64. Audoenus of Rouen, *Vitae Eligii liber II*, p. 705.12; Martin of Braga, *Canones*, 72.

65. Caesarius of Arles, *Sermo* 13.5; Audoenus of Rouen, *Vitae Eligii liber II*, p. 707.2–7; Pirmin of Reichenau, *De singulis libris canonicis*, p. 55.13–14; cf. Maximus of Turin, *Sermones*, 30, 31.

66. Audoenus of Rouen, *Vitae Eligii liber II*, p. 707.7–9.

67. Dölger, "Planetenwoche," pp. 202–206.

68. Caesarius of Arles, *Sermones*, 13.5, 19.4, 52.2; Audoenus of Rouen, *Vitae Eligii liber II*, p. 706.4–5.

69. Martin of Braga, *De correctione rusticorum*, 16. 11; Pirmin of Reichenau, *De singulis libris canonicis*, p. 54.20.

Figure 7. Calendar with planetary deities. From Altbachtal sacred precinct in Trier (ca. A.D. 275). At the top are busts of the planetary deities: Saturn (Saturday); Sol (Sunday); Luna (Monday); Mars (Tuesday); Mercury (Wednesday); Jupiter (Thursday); and Venus (Friday). Rheinisches Landesmuseum Trier. Used by permission.

or return since, as Caesarius and Eligius pointed out, all the days were made by God.[70]

Eligius's most revealing condemnation is of pagan solstitial solemnities with singing and dancing on St. John's day. Pagan solar customs still threatened a Christian festival that had been deliberately established over three centuries earlier so that Christian feasts would mark the solstices and equinoxes.

Christian polemics against astrology, Christian astrological heresies, and pagan astral cults reveal the same concerns that we have seen in the commentaries on the Old and New Testaments, although with varying emphases. Discussions of both philosophical and popular astrology strove to defend human freedom from astral constraints; hence, they tended to ignore the concept that the heavenly bodies follow an unchanging divinely established course, which might seem to support fatalistic determinism. Yet this concept did emerge in condemnations of propitiatory rituals at lunar eclipses, when it could counter a polytheistic notion of the stars and planets as independent agents.

Although a strictly deterministic astrology was universally condemned in these discussions, as in the commentaries the role of the stars was left somewhat open. To some the stars could influence the physical elements, to others the stars acted as signs, and some others held that with the coming of Christ and the establishment of a new dispensation, the stars no longer provided valid portents.

Running as a constant theme through Christian discussions of the stars and planets was the insistence that the heavenly bodies are created beings and that they have been subject to the Lord's decrees from the day of their creation. The Christian dispute with astrology was not merely with pagan superstition, it was also a dispute over a concept of the universe. To the pagan or the Manichee, the world was an arena of conflict among opposing forces; to the Christian all creation was ruled by a single supreme Lord. Taken together, these discussions reflect the continued acceptance of the heavens as an example of divinely established natural order which informed Christian astronomical practice through the turbulent era from the fourth to the tenth centuries.[71]

70. Caesarius of Arles, *Sermones*, 54.1, 193.4; Audoenus of Rouen, *Vitae Eligii liber II*, p. 705.10–12; Martin of Braga, *De correctione rusticorum*, 16.11–12; Pirmin of Reichenau, *De singulis libris canonicis*, pp. 54.20–55.1.

71. This continuing belief in the rationality of the cosmos is ignored by those who would see its sudden emergence as marking a revolutionary change in Western thought. See, e.g., Tina Stiefel, "Heresy of Science."

PART TWO

# *The cultivation of early medieval astronomies*

# CHAPTER FOUR

# Continuity and change in solar ritual

And who can count the different crowds and numberless peoples flocking
    from all the provinces
some for the abundant feasting, others for the healing of their afflictions,
    others to watch the pageant of the crowds, others with great gifts
    and offerings
to join in the solemn celebration of the feast of Saint Brigit . . . on the first
    day of the month of February.

Cogitosus, Sanctae Brigidae vita (ca. 650)[1]

The most neglected part of early medieval astronomy concerns the astronomical beliefs and practices that had developed north of the Mediterranean basin before Caesar and his successors incorporated Gaul and Germania into the Roman Empire.
This neglect stems from historians' focus on mathematical astronomy and philosophical cosmology, which tends to ignore the practical concerns that drove much
early study of the heavens. It leads to a misleading historiography in which medieval
culture is reduced to a blank photographic plate that passively absorbed a new image
of the heavens upon its first exposure to Greek and Arabic astronomy. Rather, medieval culture was more like a sketch pad already bearing the outlines of indigenous
astronomies, outlines that would be partly erased and partly incorporated through a
creative process to define the contours of a new astronomical picture.

We have already seen traces of this earlier astronomy in the stone monuments
of prehistoric Europe and in the reactions of Christian apologists to survivals of
indigenous astral cults. We have few accounts, however, of the content of these
astronomical traditions, of their roles and significance within their cultural context, or of the people who developed and used them. Archaeological evidence,
such as astronomical alignments and cult objects representing the Sun, the Moon,
and other astral divinities, indicates a concern with the heavens. But archaeological
evidence, by its very nature, cannot tell us much about the development, significance, and practice of these astronomical ideas.

A few written sources do provide fleeting glimpses of the practitioners of these
traditional astronomies. These descriptions, however, stem from a limited number
of Greek and Roman writers who were not concerned with subtle distinctions

---

1. Cogitosus, *Sanctae Brigidae vita*, 32.10, transl. S. Connolly and J.-M. Picard, p. 27.

among the ideas and practices of their barbarian Germanic and Celtic neighbors. Later writings, especially those from Ireland, provide a native, if Christian, view of Celtic practices, but do so at a time when these traditions were already in decline.

Frequently appearing in the written sources describing the Celtic societies of Gaul, Britain, and Ireland are three privileged scholarly classes: bards, *filid* (or *fáith*), and druids. The bards are universally described as poets; the *filid*, variously as prophets, seers, and keepers of genealogical, historical, and geographic lore;[2] and the druids seem most like what we would expect of astronomer-priests.

Beginning with the Greek and Roman ethnographers, we have been presented with two opposing views of the druids. Some writers were openly sympathetic to their "barbarian" neighbors, depicting the druids as philosophers, concerned with the study of ethics and nature, and even tracing the origins of philosophy to the barbarians. Others saw the druids as uncivilized, describing druid priests engaged in barbaric activities extending even to human sacrifice.[3] Recent studies echo these discordant perceptions.

Chadwick, from her examination of classical texts, concluded that the druids formed a scholarly group given to the study of nature (*physiologia*), ethics, philosophy, and theology. Chadwick argues that these descriptions do not reflect "the characteristics of priests, but of philosophers – and, especially, I would add, of Greek philosophers." Ross's synthesis of classical texts and archaeological evidence led her to conclude that the druids, "far from being the sophisticated philosophers that the Romans . . . envisaged them to be, were but little different from the priest/shamans of the entire barbarian and later pagan world."[4]

Is there any way to reconcile, at least in part, these diametrically opposed interpretations? This reconciliation might be found by asking what a Greek or Roman author, especially one given to sympathetic portrayal of his barbarian neighbors or one seeking the primitive origins of Greek philosophy, would see in the druids that could be called philosophy.

One answer might be found in the tradition stemming from the lost *De symbolis Pythagoricis* of Alexander Polyhistor (fl. ca. 50 B.C.), which sought to place the roots of Greek philosophy among the barbarians. Polyhistor claimed that philosophy had flourished before the Greeks among the prophets of Egypt, the Chaldeans of Assyria, the druids of Gaul, the shamans of Bactria, the *philosophati* of the Celts, and the Magi of Persia.[5] In both the Chaldeans and Magi we see known

---

2. On the *filid* as historians, see Hughes, *Early Christian Ireland*, pp. 165–166; Kenney, *Sources for the History of Ireland*, vol. 1, pp. 2–5.

3. N. Chadwick, *The Druids*, provides an excellent discussion of classical texts on the druids. Nonetheless, she fails to convince the reader of her central thesis that druidism was "a provincial peripheral survival" (p. 102) of Greek philosophy.

4. N. Chadwick, *The Druids*, p. 100; Ross, *Pagan Celtic Britain*, p. 61.

5. N. Chadwick, *The Druids*, pp. 58–68. Among later writers reflecting Polyhistor's views Chadwick cites Dio Chrysostom, *Oratio*, 49.8.6; Hippolytus, *Philosophumena*, 1.22; Clement of Alexandria, *Stromata* 1.15.70–1, *PG* 8, col. 778; and Cyril of Alexandria, *Contra Julianum*, 4, *PG* 9.

representatives of astral religion, reinforcing the picture of the druids found in later accounts.

These descriptions associate the practice of two crafts with the druids: Pliny noted the importance of the practice of medicine, and Julius Caesar and Pomponius Mela, both drawing on the lost work of the Stoic philosopher and historian Posidonius (ca. 135 B.C.–ca. 50 B.C.), noted the study of the stars and their motions.[6] And here, of course, we see the two crucial ingredients in the origins of the Greek study of nature. The study of the heavens and the art of medicine have provided historians, from Plato to the present, with keys to the origins of Greek science.[7] If astronomy and medicine correspond to the sympathetic classical descriptions of druids as possessing the origins of philosophy, they also form much of the knowledge of nature revealed by modern ethnographic studies of tribal societies.

Later Irish writers, considering a declining druidism from a Christian perspective, saw divination playing a much greater part. Ross has noted a number of Irish texts discussing the interpretation of dreams and of such omens as the flight of birds, albeit not explicitly tying such interpretation to druids.[8] The druids themselves watch the skies for signs in a pair of Irish scriptural commentaries on the coming of the three kings. Sedulius Scottus, writing in Latin in the ninth century, attributes the following words to Joseph on seeing the three kings: "It seems to me that these who are coming are soothsayers (*augures*). Notice how every moment they look up to heaven, and argue amongst themselves." The Irish *Lebar Brecc*, a fifteenth-century compilation of earlier material, describes the Magi as three druids to whom the kingly star appeared. In a clear allusion to the same tradition used by Sedulius, *Lebar Brecc* has Joseph comment: "I fancy . . . that it is the omen art of the druids, and it is soothsaying that they are practising, for they take not a single step without looking up. . . ."[9] In these Irish texts we see again the kind of observations of nature, of the course of stars and the passage of birds, associated ethnographically with soothsayers and keepers of the calendar.

None of the descriptions of the druids imply scientific practices divorced from religious activities. Rather, they suggest practitioners of an astral cult, calendar keepers, and powerful figures in the community. The healer and the Sun watcher possess knowledge that is both sacred and powerful; by knowing the course of disease or of the seasons one can – in a sense – foretell the future. Among "primitive" peoples – and we must recall that the Greeks and Romans were not that far from their primitive pasts – the knowledge of nature and the knowledge of

6.  Pliny, *Hist. nat.*, 24.62, 29.5; N. Chadwick, *The Druids*, pp. 31–32; Caesar, *De bello Gallico*, 6.14; Pomponius Mela, *De chorographia*, 3.2.18–19.
7.  Plato, *Timaeus*, 47A; G. E. R. Lloyd, *Magic, Reason, and Experience* (Cambridge: Cambridge Univ. Pr., 1979).
8.  Ross, *Pagan Celtic Britain*, pp. 259–261.
9.  Cited in Robert E. McNally, "The Three Holy Kings in Early Irish Writing," in P. Granfield and J. A. Jungmann, eds., *Kyriakon: Festschrift Johannes Quasten*, vol. 2, pp. 667–690 (Münster i. W.: Aschendorff, 1970), here pp. 685, 688.

the sacred went hand in hand. Thus any Greek or Roman ethnographer who wished to affirm the wisdom of the barbarians could discern *physiologia* or *philosophia* in the practices of barbarian wise men. It is the character of these barbarians' understandings of the heavens that we must now try to discern.

## Calendar and ritual in Celtic Gaul

The druids, and their knowledge of nature, did not vanish abruptly with the arrival of the Romans; druids and their descendents continued to hold leading scholarly positions in Gaul. Ausonius (ca. 310–ca. 394) tells of three professors at Bordeaux, son, father, and grandfather, who came from a family of noted druids, the eldest of whom, Belenus, had been a priest at a temple of the Celtic Apollo.[10] We find explicit written evidence for Celtic astronomical practice in the second-century Calendar of Coligny, well before the final dissolution of Celtic culture in Gaul. This monumental bronze tablet (1.48 × 0.90 meters), found in fragments some eighty kilometers north of Lyons, is the most extensive sample of an early written Celtic language.[11]

As reconstructed, the calendar covers a period of five years, totaling 1835 days, with complete entries for 640 days and partial entries for an additional 433 days (Figs. 8 and 9). There are 62 lunar months in these five years, with their Celtic names written in the Latin alphabet. Small sockets into which pegs could be inserted had been drilled to the left of each day, most likely to mark the current day.

Like much of Gallic culture, the Calendar of Coligny combines native Celtic elements with ideas drawn from centuries of contact with Romans and Greeks. It is an excellent example of the kind of inscribed calendar that had developed in Mediterranean city-states, yet its content is clearly indigenous. The earliest such Greek calendars described the appearance and disappearance of stars in the course of an entire solar year, with sockets drilled next to each day into which pegs marking the current day, and perhaps the changing phases of the Moon, could be inserted. Roman examples were structured on the solar year but variously added places to mark the current day in the lunar or calendar month, in the week or in the Roman eight-day market week, and the place of the Sun or Moon in the zodiac (Fig. 10).[12]

10.  Ausonius, *Commemoratio professorum Burdicalensium*, 4, 5, 10.21–29.

11.  Duval and Pinault, RIG, 3, les calendriers; E. MacNeill, "Calendar of Coligny." A few fragments of a similar calendar were found near a Gallo-Roman sanctuary at Villards d'Héiria, some 30 km east of Coligny.

On the calendar's date, see Duval and Pinault, RIG, 3, pp. 24, 26, 35–37; cf. E. MacNeill, "Calendar of Coligny," pp. 4–7; Proinsias MacCana, "Celtic Religion."

12.  Albert Rehm, in *Paulys Real-Encyclopädie der classischen Altertumswissenschaft*, 18, 2, s. v. Parapegma, cols. 1295–1302, 1361–1366; *Corpus Inscriptionum Latinarum*, I, pp. 218, 253; Eriksson, *Wochentagsgötter*, pp. 17–37; Neugebauer, *HAMA*, pp. 587–589.

The later simplified calendars, which note only the days of the week and the four seasons, were not monumental public calendars, but small mass-produced terra-cotta castings.

Figure 8. Schema of the Calendar of Coligny (after Daviet). Listing of the months and their durations as arranged on the calendar.

The months and their durations (in days), by year:

| Month | YEAR 1 (385 DAYS) | YEAR 2 (355 DAYS) | YEAR 3 (385 DAYS) | YEAR 4 (355 DAYS) | YEAR 5 (355 DAYS) |
|---|---|---|---|---|---|
| IN¹ | 30 | | | | |
| SAM | 30 | 30 | 30 | 30 | 30 |
| DUM | 29 | 29 | 29 | 29 | 29 |
| RIV | 30 | 30 | 30 | 30 | 30 |
| ANA | 29 | 29 | 29 | 29 | 29 |
| IN² | | | 30 | | |
| OGR | 30 | 30 | 30 | 30 | 30 |
| CUT | 30 | 30 | 30 | 30 | 30 |
| GIA | 29 | 29 | 29 | 29 | 29 |
| SIM | 30 | 30 | 30 | 30 | 30 |
| EQU | 30 | 30 | 30 | 30 | 30 |
| ELE | 29 | 29 | 29 | 29 | 29 |
| EDR | 30 | 30 | 30 | 30 | 30 |
| CAN | 29 | 29 | 29 | 29 | 29 |

Abbreviations (an asterisk marks unattested forms):

| | | | |
|---|---|---|---|
| In | Intercalary | Ana | Anagantio- |
| Sam | Samon- | Ogr | *Ogronn- |
| Dum | Dumann- | Cut | Cutios |
| Riv | Rivros | Gia | *Giamoni- |
| Sim | *Simivisonna- | | |
| Equ | Equos | Edr | *Edrini- |
| Ele | Elembiu | Can | Cantlos |

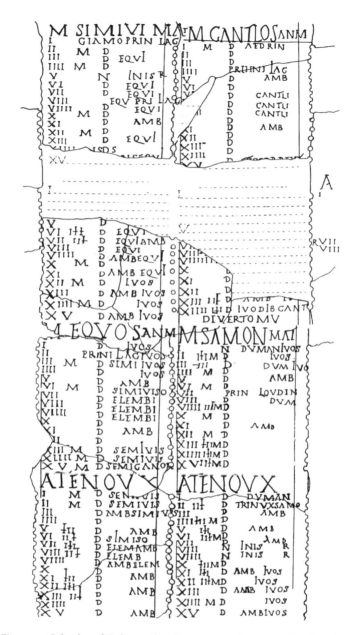

Figure 9. Calendar of Coligny (detail). Late second century. Reading down the first column are the months SIMIVI [SONNA] and EQVOS; in the second column are CANTLOS of the first year and SAMON of the second. Each month is divided by the heading ATENOVX. Note "PRINI LAG IVOS" at 2 EQVOS, "PRINNI LAG" at 4 CANTLOS, and "TRINUX SAMO [NI SINDIV]" at 2ª SAMON. Reprinted from P.-M. Duval and G. Pinault, *Recueil des inscriptions gauloises*, vol. 3, les calendriers. Used by permission.

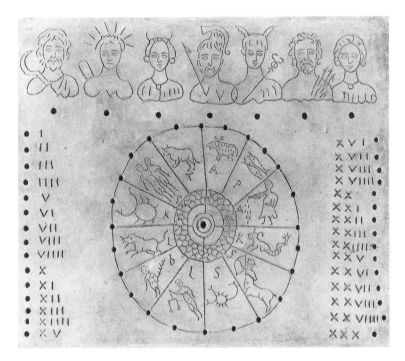

Figure 10. Calendar from a Christian shrine. Terra-cotta copy of a *steck-kalender* from the oratory of St. Felicity, Rome, fourth century. The calendar contains sockets in which markers could be inserted to indicate the day of the month (numbered one through thirty), the place of the Sun in the zodiac, and the day of the week by the corresponding planetary deity. The central four circles below the zodiac represent the spheres of the four elements. Martin-von-Wagner Museum der Universität Würzburg, inv. no. G14. Used by permission.

The Coligny calendar is primarily lunar, with five months of twenty-nine days and seven months of thirty days. Two additional intercalary months are added every five years, producing two full years of thirteen months (385 days) and three hollow years of twelve months (355 days). Over the five years of the inscription, the average length of the year is 367 days (see Fig. 8).[13] Signs of the earlier existence of this five-year luni-solar calendar appear in Greek and Roman de-

---

13. In this discussion I follow the extant fragments rather than the hypothetical reconstructions of E. MacNeill, "Calendar of Coligny," pp. 27–29, and G. Pinault, pp. 411–417, in Duval and Pinault, RIG, 3.

scriptions of the Gauls. Diodorus Siculus (fl. 60 B.C.–21 B.C.) reports that the Gauls sacrificed prisoners to the gods every five years.[14]

The months and days in the Calendar of Coligny are not simply undifferentiated units of time. The thirty-day months are generally described as *matu-*, that is, full, complete, or good, and the twenty-nine-day months are *anmatu-*, partial, incomplete, unpropitious.[15] The same characteristics are applied to the days as well as the months; selected days of the complete months are also marked *matu-*, reinforcing those days' propitious nature. Only one of the dates in the calendar can be associated with a festival. The entry *★trinox Samoni sindiu*, on the seventeenth day of the lunar month *Samon-*, which Duval translates as "this day, the three nights of Samhain," may represent either the Celtic mid-quarter feast of Samhain or the date of the summer solstice.[16]

The calendar divides both months and years into two halves. The days of the first half of the month are numbered from one to fifteen; then the count starts over, running from one to fourteen or fifteen depending on whether the month has twenty-nine or thirty days. The intercalary months reflect a similar division of the year into halves. The first intercalary month falls directly before the first in the calendar's regular sequence of months, *Samon-*, Summer month, which has been variously taken to mark the beginning, middle, or end of *samh*, or summer; the second intercalary month falls two and a half years later, directly before *★Giamoni-*, Winter month, taken as the beginning, middle, or end of *gamh*, or winter.

The intercalary months reflect a concern to reconcile the solar year and the lunar calendar. Yet the length of the calendar's year averages 367 days, an error of almost two days per year. Without adjustment, in five years the calendar will gain four days on the lunar month and nine days on the tropical year.[17] Shortening some of the months would bring the calendar closer to the lunar phases, but the calendar would still be out of synchronization with the solar year. The cycle attested by the calendar provides for 62 months in five years, or 372 months in thirty years, but thirty tropical years correspond to only 371 lunar months.[18] As it stands, the calendar is clearly too approximate to be continued indefinitely without further adjustment; some method of observation or computation must

14. Diodorus Siculus, *History*, 5.32.6.

15. A significant exception is the month *Equos*, which while thirty days long, is labeled as *anmat* like the shorter months. E. MacNeill, "Calendar of Coligny," pp. 27–29, and following him, G. Pinault, pp. 411–417, in Duval and Pinault, RIG, 3, suggested that the month *Equos* was originally twenty-nine days long, but by the time of the Calendar of Coligny alternated between a length of thirty days in the first, third, and fifth years (attested for the first and fifth) and twenty-eight days in the second and fourth years.

16. Duval and Pinault, RIG, 3, pp. 403, 426, 427. E. MacNeill, "Calendar of Coligny," pp. 19–21.

17. The Calendar of Coligny (1835 days); five tropical years (365.2422 × 5 = 1826.21 days); sixty-two synodic months (29.53054 × 62 = 1830.89 days).

18. Thirty tropical years (30 × 365.2422 = 10957.266 days); (10957.266 / 29.53054 = 371.048 months = 371 months, 1.4 days).

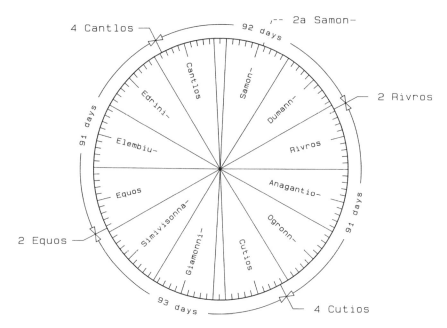

Figure 11. Seasonal feasts in the Calendar of Coligny. Tick marks every three days. The two intercalary months are represented by segments of six days; the average length of one thirty-day month every five years.

The dates 2 Rivros, 4 Cutios, 2 Equos, and 4 Cantlos divide the year into four quarters of about 91, 93, 91, and 92 days. The date 2ᵃ Samon-, falling midway between 4 Cantlos and 2 Rivros, is either mid-summer day or Samhain, the end of summer.

have been used to adjust the length of the lunar months and to regulate their place in the solar year.[19]

Further examination of the Calendar of Coligny shows that the calendar divided the year into four equal parts and marked at least one of the mid-quarter festivals. Four days sharing the uncommon notation *Prinni Lag* . . . divide the solar year into four equal portions (Fig. 11). These four days, 2 *Rivros*, 4 *Cutios*, 2 *Equos*, and 4 *Cantlos*, fall near the beginning of the third, sixth, ninth, and twelfth months in the lunar calendar.[20] There are two plausible interpretations of these quarter divisions: most likely is that they mark the mid-quarter days of the insular Celts: Samhain, Im-

19.  Fotheringham suggested a systematic change to make the Calendar of Coligny agree with the lunar month by reducing *Equos* to twenty-nine days in three of the calendar's five years; MacNeill advanced the alternative that *Equos* be reduced to twenty-eight days in two of the five years. Such hypothetical systematic changes seem unlikely. E. MacNeill, "Calendar of Coligny," pp. 27–29; Duval and Pinault, RIG, 3, pp. 411–415.

20.  McCluskey, "Solar Year in the Calendar of Coligny."

bolc, Beltaine, and Lughnasa (falling at the beginning of November, February, May, and August in the Julian calendar). In this case *trinox Samoni sindiu* marks the summer solstice. Alternatively, they mark the solstices and equinoxes, in which case *trinox Samoni sindiu* marks the mid-quarter festival of Samhain. In either event, the calendar reflects both the Celtic mid-quarter days and an attempt to reconcile solar and lunar phenomena. Whatever procedure the Gauls may in fact have used to regulate their calendar remains uncertain.

It seems most likely, considering the practices of similar peoples and of the Greeks and early Romans,[21] that the Gauls regulated their calendar by observation rather than by any mathematical system. In this case astronomer-priests, here the druids, would shorten four months every five years as necessary by direct observation of the Moon. They would also need to omit an intercalary month every thirty years, which could be done on the basis of correlated observations of the Sun and Moon at some astronomically fixed point in the solar year. The great assembly held by the Gallic Celts at the beginning of August, marking the harvest festival of Lughnasa, would have been a likely time to make such adjustments to the calendar.

### Christening the solar calendar

The solar division of the year at days falling midway between the solstices and equinoxes, indicated by prehistoric British megalithic monuments, implicit in the Calendar of Coligny and surviving in Celtic folk traditions, finds more direct expression in the later interactions of Christian and Celtic traditions, best recorded on the insular fringes of Europe. Here as in Gaul the victorious Christians, who supplanted the bards, *filid*, and druids, exploited calendric elements from Celtic paganism, a paganism they would describe as "splendid."[22]

This same division of the seasons is specifically mentioned in several early computistical tracts. Martin of Braga noted in his *De Pascha* that just as the vernal equinox divides night and day evenly, so does it divide spring into two equal halves. The mid-sixth-century *De ratione paschali* of the Irish Ps.-Anatolius has the four seasons begin midway between the conventional dates of the solstices and equinoxes in the Julian calendar: 25 March, 24 June, 24 September, and 25 December, which by that time were already some three to seven days later than the true solstices and equinoxes. This passage was cited as authoritative by the Irish author of the seventh-century *De ratione conputandi* and in the eighth century by Bede, who consequently had the seasons begin on 7 February, 9 May, 7 August,

---

21. Neugebauer, *HAMA*, pp. 616–617, 1076.
22. *Félire Oengusso*, p. 214.

and 7 November.[23] In many Irish monasteries the seasons were divided at the traditional mid-quarter days. Columbanus's monastic *Rule*, written before 615, makes the first of November the beginning of winter. The *Life of Columba*, written near the end of the seventh century by Adomnan, abbot of Iona, implies that the fifteenth day of June is the middle of summer.[24] The relations between the two sets of major divisions of the year, the solstices and equinoxes, and the days midway between them are depicted in two nearly identical ninth-century astronomical manuscripts (Fig. 12).[25]

This traditional Celtic division of the year was not mentioned just in texts. It can be identified more precisely by considering the transformation of the four mid-quarter festivals, Samhain, Imbolc, Beltaine, and Lughnasa, into Christian feasts. Just as early Christians had developed a ritual calendar, with major feasts of Christ and John the Baptist at the Julian calendar's canonical dates of the solstices and equinoxes, so did later generations of Christians transform the mid-quarter festivals into Christian feasts which took on aspects of their pagan counterparts. When a pagan festival was transformed into a Christian holy day, its date was no longer determined by traditional solar observations, but became fixed in the Julian calendar used by the Christian church. Since the Julian calendar does not exactly follow the Sun, but errs by some three days in 400 years, a particular Julian date can coincide with a particular solar date only for about two centuries. This offers an opportunity to confirm whether a Christian feast that folklore associates with a mid-quarter day actually corresponded with the observational solar calendar at the time it was established.[26]

Considering the historical development of the Christian liturgical calendar, by investigating a group of Christian feasts that folklore and tradition associate with one of the mid-quarter festivals, we find that at the time of their establishment these feasts actually fell within a few days of that solar mid-quarter festival.[27] I

23. Martin of Braga, *De pascha*, 5.67–77; ps.-Anatolius, *De ratione paschali* 14, in Krusch, *84jährige Ostercyclus*, p. 327; *De ratione conputandi*, 48; Bede, *De temporum ratione*, 35.35–55.

24. Columbanus, *Regula Monachorum*, 7, p. 130.2, in Columbanus, *opera*; Adomnan, *Life of Columba*, 2.2, pp. 119, 326–329. Adomnan's date is closer to the actual date of 18–19 June than was the conventional date, although it cannot be taken as mathematically precise.

25. Munich, Bayerische Staatsbibliothek, CLM 210, fol. 136v; Vienna, Österreichische National-bibliothek, Cod. 387, fol. 137r.

26. This comparison can provide only a general indication of a match between an observational solar calendar and the Julian calendar at the time the feast was established.

   Considering the ¾-day variation of the Julian year introduced by leap years, a given day in the Julian calendar can vary by 1¾ days in solar time. This allows a leeway of about 225 years within which a Julian date and a solar date sometimes coincide exactly. Furthermore, ethnographic evidence suggests that observational solar calendars have errors of one or two days. Given these errors the data are not sufficiently sensitive to provide a clear distinction between the two different models for dividing the solar year, which typically differ by only two days.

27. This topic is treated in greater detail in McCluskey, "Mid-Quarter Days."

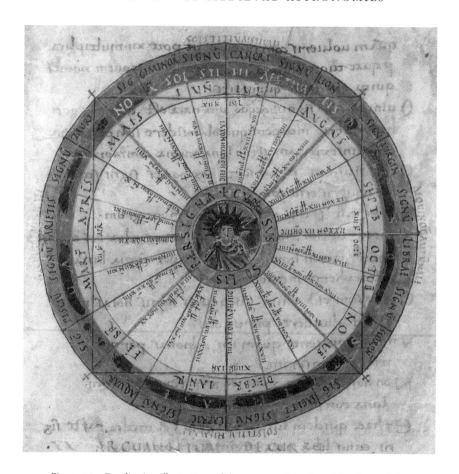

Figure 12. Carolingian illustration of the course of the Sun. The dates of the solstices and equinoxes are distinctly labeled, and those midway between the solstices and equinoxes are marked by crosses at the corners of the square. Vienna, Österreichischen Nationalbibliothek, MS 387, fol. 137r. Used by permission.

have excluded from this discussion feasts that originated in the Mediterranean region and feasts that other evidence justifies excluding, even if later folklore associated them with the mid-quarter days.[28] The characteristics of the feasts considered are summarized in Table 1.[29]

28. Among the feasts excluded are the Purification (Candlemas, 2 Feb.); St. Blaise (3 Feb.), a feast which became popular only in the fourteenth century; Sts. Philip and James (1 May); St. James the Greater (25 July); St. Christopher (25 July); St. Anne (26 July), a late fourteenth-century feast; St. Pantaleon (28 July); St. Abdo (30 July); St. Peter's Chains (1 Aug.); All Souls (2 Nov.);

Table 1. *Relation to Christian feasts in the Julian calendar to Celtic festivals in the solar calendar*

| Festival | Year Christian feast was established | Computed years solar festival fell on Christian feast | |
| | | Equal-time model | Equal-angle model |
| --- | --- | --- | --- |
| *Samhain* | | | |
| All Saints (1 Nov.) | 798 | 767–996 | 567–786 |
| *Imbolc* | | | |
| St. Brigit (1 Feb.) | ca. 525–ca. 650 | 678–908 | 416–660 |
| *Beltaine* | | | |
| Conception BVM (2 May) | 902 | 817–1046 | 641–857 |
| Conception BVM (3 May) | ca. 800 | 686–915 | 519–734 |
| *Lughnasa* | | | |
| St. Justus (4 Aug.) | ca. 390–467 | 464–695 | 711–925 |
| St. Oswald (5 Aug.) | 642–ca. 685 | 332–564 | 588–803 |

The feast of All Saints (1 Nov.) is the feast most widely recognized as a successor to a mid-quarter festival, in this case Samhain. Samhain means the end of summer and marks one of the two primary turning points of the Celtic year; the festival concerns the bonds between the living and the dead, a point it shares with its

and St. Martin (11 Nov.), a feast which assumed its calendric significance only after the Gregorian calendar reform.

29.  The motion of the Sun was computed using the parameters in Tuckerman, *Planetary Positions*, vol. 56, pp. 8–15, vol. 59, p. 13, verified by spot checks against his values. The longitude was taken as that of Greenwich, rather than Tuckerman's Babylon meridian. Given Tuckerman's estimates of errors, the computational errors in solar longitude are insignificant and the largest possible error due to uncertainty in the earth's rotation is certainly less than the 0.1° which he estimates for 601 B.C. These position errors can produce an error in the tabulated years of as much as twenty years.

Christian successor.[30] Despite the occasional challenges by historians of liturgy, there is little doubt that All Saints and Samhain have long been closely related.

The establishment of the feast can be firmly dated to the end of the eighth century. All Saints is mentioned in early martyrologies, including the eighth-century martyrology attributed to Bede, and in the first half of the ninth century it emerged as a universal feast, in which all the saints of the whole world would be commemorated everywhere in one solemn celebration.[31] Alcuin of York, a leading figure in the literary and astronomical revival in Charlemagne's court, brought this observance from England to the continent. He composed several masses for the feast and vigil of All Saints and in 798 drew upon the conventional theme of illumination to recommend the feast to his friend, the missionary Bishop Arno of Salzburg, for its opposition to paganism. Since all the saints of the New Testament "can shut heaven to the unbelieving and open it to the faithful," this most solemn feast should be preceded by three days of prayers, fasting, and alms-giving so that Christ, the true light, would illuminate the hearts of believers. Alcuin juxtaposed Christ, the true light, to the pagan feast's theme of death at a season of growing darkness.[32] In 802 the Bavarian synod of Riesbach, presided over by Arno, listed All Saints among the church's important feasts. Later in the century Pope Gregory IV urged Emperor Louis the Pious to decree that the feast be celebrated in France and Germany on November first.[33] Around the year 800 the solar date of Samhain fell either on the first or second of November, using the arithmetical division of the year into parts having an equal number of days (the equal-time model in Table 1)[34] or on the third of November, using the geometrical division of the zodiac into equal angles, which would place the Sun midway between the autumnal equinox and the winter solstice.

The feast of St. Brigit (1 Feb.) is much less a rival to the pagan feast of Imbolc that it replaced, than it is a Christian adoption of pagan traditions. The name and attributes of St. Brigit continue those of a Celtic triad of divine sisters, all named Brigit. The chief of them was honored by the *filid* as patroness of poetry; her sisters were patronesses of medicine and of metalwork.[35] The chief Brigit was

30.  De Vries, *Keltische Religion*, pp. 229–230; McCluskey, "Mid-Quarter Days," pp. S4– S6.

31.  Ps.-Bede, *Homiliae subditae*, LXX (*In eadem solemnitate Omnium Sanctorum*); John Hennig, "The Meaning of All the Saints," *Mediaeval Studies*, 10(1948):147–161.

32.  Alcuin, *Epistola* 193, *MGH*, Epp., 4, p. 321.

33.  *MGH*, Concilia, 2, pars. 1, p. 197; Jean Deshusses, "Les Anciens Sacramentaires de Tours," *Revue Bénédictine*, 89(1979):281–302, here pp. 291–295; André Wilmart, "Un Témoin Anglo-Saxon du Calendrier Metrique d'York," *Revue Bénédictine*, 46(1934):41–69, here pp. 51–55; *Mar-tyrologe d'Adon*, p. 371. I follow the dates given by Deshusses on the basis of more recent research.

34.  The equal-time model reflects the observational practice identified in studies of the alignments of megalithic monuments. Thom, *Megalithic Sites in Britain*, pp. 102, 107–117.

35.  *Sanas Chormaic*, p. 23.

herself the mother of another divine triad: Brían, Iuchar, and Uar.[36] As Brig she is mentioned in the Martyrology of Oengus on the day before St. Brigit's feast.

The Irish Brigit has been identified with the Gaulish Minerva, named by Julius Caesar as the only goddess among the five principal gods of the Gauls, who worshiped her as author of the arts and crafts.[37] De Vries sees hints of solar cult in dedications to Minerva Belisma (St. Lizier, France) and Minerva Sulis (Bath, England), which he suggests are pre-Celtic survivals.[38]

St. Brigit, like her pagan precursor, was specially associated with the arts of women and with fertility; she was a milkmaid, and she was described variously in popular lore as the midwife at Christ's birth and as taking the child to her breast. There are even extravagant Irish claims that St. Brigit was the mother of Christ, which may continue the goddess Brigit's role as mother of gods.[39] Her cult, the most popular of any Irish saint, spread rapidly through Britain and the continent and is found in sources as diverse as early martyrologies and litanies and nineteenth-century Scottish folklore. The rapid and wide diffusion of her cult may reflect the successful replacement of a popular pagan cult with special appeal to women by a Christian cult reflecting their concerns.[40]

Traces both of Brigit, the bright one, and of a solar cult are seen in a seventh-century hymn to St. Brigit, which begins:

> Brigit ever excellent woman,
> golden sparkling flame,
> lead us to the eternal kingdom,
> the dazzling resplendent Sun.[41]

There are other suggestions of St. Brigit's special relationship to the Sun. In one recurring legend in the *Lives* of St. Brigit, she hung her rain-soaked cloak on a sunbeam, which, rather than let the cloak fall to the ground, continued to illuminate the surrounding countryside well into the night.[42]

The date of the establishment of Brigit's feast can only be approximately defined. Brigit's death is assigned various dates around the year 525; a great annual

36.  De Vries, *Keltische Religion*, p. 143.
37.  Caesar, *De bello Gallico*, 6.17.
38.  De Vries, *Keltische Religion*, pp. 78–79, 132–133.
39.  Gougaud, *Saints irlandais hors d'Irlande*, p. 193; Kenney, *Sources for the History of Ireland*, vol. 1, pp. 356–358; *Félire Oengusso*, pp. 39, 58.
40.  Bowen, "Cult of St. Brigit." A survey of Irish saints in 21 early British and continental litanies shows Brigit in 20; Columban in 14, and Patrick in 9. Brigit is also the most frequently named saint in Alexander Carmichael's *Carmina Gaedelica* (Edinburgh: Oliver & Boyd, 1926). Gougaud, *Saints irlandais*, pp. 18–19, 189–192.
41.  *Ultan's Hymn*, pp. xxxviii, 323–326.
42.  *Acta Sanctorum*, Febr. I, 1, pp. 131, 136, 142, 161.

assembly on her feast day is described in Cogitosus's *Life of Saint Brigit*, which was written shortly after 650.[43] Within this range of dates from 525 to 650, the solar date of Imbolc fell either on 2–3 February using the division of the year into equal units of time or between 31 January and 2 February according to the geometrical division of the zodiac by equal angles.

No well-established Christian feast can be clearly associated with the pagan festival of Beltaine. Yet a number of early and modern texts indicate the connection of the Blessed Virgin with Beltaine, May Day, and the crowning of a May Queen as a symbol of fertility. Although the official church could not sanction a major Marian feast in Eastertide, a popular rite arose in which Mary was crowned as May Queen, and the month of May was dedicated to her. As late as the nineteenth century, a traditional Beltaine blessing from the Hebrides invoked the protection of the Blessed Virgin.[44]

Signs of a May feast of Mary appear in a few early sources from the British Isles. The *Martyrology of Oengus* (ca. 800) marks the third of May as a "great feast of the Virgin Mary," and the *Martyrology of Tallaght* (ca. 800) designates that date as celebrating the conception of the Virgin Mary. In the metrical Calendar of Hampson (soon after 902) the date of the feast of the Conception of the Virgin has moved to the second of May.[45] Around the year 800, the solar date of Beltaine fell on 3 May according to the equal-time model and between 1–2 May according to the equal-angle model; around the year 902, the dates had shifted to 2–3 May according to the equal-time model and 30 April to 1 May according to the equal-angle model. The correspondence between the changing dates of the feast provided by the texts, and the changing date of Beltaine defined by the movement of the Sun, suggests the assimilation of a Christian feast into the traditional calendar. In this case, however, since the feast was not widely incorporated into the liturgical calendar, it did not become fixed in the Julian calendar but continued to follow the appearances of the Sun.

One of the most thoroughly studied of the mid-quarter festivals is Lughnasa, a harvest assembly and festival dedicated to the god Lugh, "the shining one," skilled in all the arts and sharing elements of the Roman Mercury as bringer of plenty, Mars as god of war, and even the solar Apollo. Lugh's festival is connected with

43. Sean Connolly and Jean-Michel Picard, "Cogitosus's *Life of Brigit*," *Journal of the Royal Society of Antiquaries of Ireland*, 117(1987)5–27; cf. Richard Sharpe, *"Vitae S. Brigitae,"* p. 87, n. 3.

44. Marina Warner, *Alone of All Her Sex: The Myth and the Cult of the Virgin Mary* (New York: Alfred A. Knopf, 1976), pp. 281–283; Carmichael, *Carmina Gaedelica*, vol. 1, p. 187.

45. Bishop, "Feast of the Conception"; *Félire Oengusso*, p. 122; *Martyrology of Tallaght*, p. 39; McGurk, "Calendar of Hampson."

  The Bollandist Paul Grosjean considers the evidence for an early Marian feast in May to be inadequate; he maintains that these isolated testimonies arose from a combination of scribal errors. Grosjean, "Prétendue fête de la Conception."

assemblies, often held at hilltop shrines, throughout the Celtic fringe of pre-Christian and modern Europe.[46] In many cases folklore, historical accounts, and local traditions connect the celebration with a Christian feast day. Two local feasts provide further details for the transformation of the traditional solar calendar, the feasts of St. Oswald (5 Aug.) and St. Justus of Lyons (4 Aug.).

The cult of St. Oswald, king of Northumbria, is the prototype of a series of cults of royal saints springing from the contact of Anglo-Saxon and Celtic paganism with Christianity.[47] The Anglo-Saxon Oswald, who had been converted by Irish monks while in exile in Scotland, returned to take the crown, lead his people to Christianity, and reunify his fragmented kingdom. Oswald's feast commemorates his death in battle on 5 August 642 at the hands of Penda, the pagan king of Mercia. Soon many cures at the battlefield were attributed to the martyred king.[48]

Oswald's successor, Oswiu, sponsored this dynastic cult, a point to which I will return later. In the last quarter of the century Oswiu's sister, Osthryth, who was also Oswald's niece and the wife of King Ethelred of Mercia, moved Oswald's body from the battlefield to the monastery of Bardney in disputed territory bordering the two realms. Oswiu continued to develop Oswald's cult by transferring his potent relics to the center of the kingdom: the king's miraculously preserved right hand and arm to the Church of St. Peter in the royal city and mercantile center of Bamburgh and his head to Bishop Aidan's monastery on the nearby island of Lindisfarne. Bede tells us that around 685

not only in this monastery [Selsey] but in many other places, the heavenly birthday of this king and soldier of Christ began to be observed yearly by the celebration of masses.[49]

Within half a century a flourishing dynastic cult had grown up around St. Oswald, which, like Brigit's, spread far beyond its local origins. His feast on 5 August was frequently mentioned in the martyrologies of English churches; some fifty churches (chiefly in the north of England) were dedicated to the martyred king.[50]

46. M. MacNeill, *Lughnasa*, pp. 1–11; Daniel F. Melia, "The *Grande Troménie* at Locronon: A Major Breton Lughnasa Celebration," *Journal of American Folklore*, 91(1978):528–542; MacCana, *Celtic Mythology*, pp. 24–25.

47. For two different perspectives, compare Ridyard, *Royal Saints*, with Chaney, *Cult of Kingship*.

48. Bede, *Hist. eccl.*, 3.1–3, 5–6, 9–13. Bede's account is repeated in Alcuin, *Saints of York*, 234–505, and Aelfric, *Lives of the Saints*. There is some question whether 5 August is the historical date of Oswald's death or the date chosen for its annual commemoration. Bede based his date on the liturgical practices at Hexham a generation after Oswald's death. Molly Miller, "The Dates of Deira," *Anglo-Saxon England*, 8(1979):35–61.

49. Bede, *Hist. eccl.*, 4.14.

50. Bede, *Hist. eccl.*, 3.6, 11, 13; Folz, "Saint Oswald Roi de Northumbrie; Folz, *Saints rois du Moyen Age*, pp. 46–48, 173–175, 182–184; Alcuin, *Saints of York*, 455–458; Frances Arnold-Forster, *Studies in Church Dedications* (London: Skeffington, 1899).

As early as the time of Bede, the cult had "spread its rays across the ocean, enlightening the peoples of Germany."[51]

To Bede and later historians Oswald typified the emerging ideal of a royal saint: a benevolent and charitable ruler and a patron and defender of the church. Yet there are also pagan elements in his story. Penda had Oswald's head and arm cut from his body and displayed atop stakes at the battlefield, which suggests either a sacrifice to Woden or the Celtic cult of the head. In a mid-twelfth-century account, Oswiu was aided in identifying St. Oswald's arm by a raven, who carried it off and dropped it on a stone from which water flowed. The raven was sacred both to the Germanic cult of Woden and to the Celtic cult of Lugh.[52] Equally striking is the solar symbolism associated with the translation of his body to Bardney Abbey in Lindsey. The monks objected to admitting the northern king's bones, which were left outside the monastery in a tent. That night a column of light, which Aelfric compared to a sunbeam, shone throughout the night above the tent, convincing all of the power of the martyred king's relics.[53]

This solar symbolism and the evidence connecting Oswald's cult to the festival of Lughnasa are confirmed by the dates of Lughnasa during the the period of the establishment of Oswald's cult. Between 642 and 685 Lughnasa fell on 3–4 August according to the arithmetical model of equal division of the year, and on 5–6 August according to the geometric model of equal division of the zodiac.

Moving from the British Isles, we find at Lyons in southern Gaul a Lughnasa festival with a well-established pedigree. Lyons lies where the Saône joins the Rhône, astride the principal trade route from Gaul to the Mediterranean, where there had long been a center of Celtic culture with a temple dedicated to Lugh. A Roman city was founded in 44 B.C. as *colonia copia Lugdunum*, the fruitful colony of the place of Lugh, which soon became a center of trade and Roman administration in Gaul. It was also the focal point of the amalgamation of Roman and Celtic culture, as exemplified by the Calendar of Coligny and the continuation under Roman auspices of a great assembly drawing representatives from all parts of Gaul.

A late fourth-century bishop of the city, Justus, gave rise to a Christian cult that continued the city's traditional rituals devoted to Lugh. Justus died about the year 390 in monastic retirement in Egypt. His remains were subsequently returned to Lyons on 4 August and buried in the cemeterial basilica dedicated to the Machabees, whose feast falls on 1 August. Sometime before 468, Sidonius Apollinaris described a great gathering of layfolk and clergy before dawn on St. Justus's feast.[54]

---

51.   Bede, *Hist. eccl.*, 3.13.
52.   Reginald of Coldingham, *Vita S. Oswaldi*, 1.17–18; Chaney, *Cult of Kingship*, pp. 117–119.
53.   Aelfric, *Lives of the Saints*, 26.176–193; Bede, *Hist. eccl.*, 3.11.
54.   *Bibliotheca Hagiographica Latina*, 4599–4600; Sidonius Apollinaris, *Epistulae*, 5.17.3–4. Sidonius does

The fifth-century internment of the saint's body in his basilica at Lyons was given solar symbolism; the anonymous *Life of St. Justus* compares it to the setting of the Sun. The *Life* also tells us that when the saint fasted in the desert, he was fed by crows like the prophet Elias. But besides this biblical connection the crow was also sacred to the god Lugh and was associated with the original Celtic foundation of the city of Lugdunum, appearing with Lugh on second-century medallions honoring the Roman city.[55]

Here again, the date of the feast of St. Justus at the time of its establishment falls within the range of computed solar dates, which are 4–6 August according to the equal-time model and 6–7 August according to the equal-angle model.[56]

At the time that the five Christian feasts just discussed were established, their dates all coincided, within the tolerance expected of horizon observations, with the dates of the four mid-quarter festivals. This agreement can best be explained by some continuation of traditional observations of the Sun during the period from the fourth to the tenth centuries in which these feasts were established.[57] In the emergence of these Christian feasts we can see the replacement of traditional mid-quarter festivals fixed in an observational solar calendar by feasts dedicated to the saints fixed in the Julian calendar. As these Christian feasts emerged, they continued many of the social functions of the pagan festivals that they supplanted. Thus by examining the cultural context in which these Christian feasts emerged, and the social roles that they played, we can gain insights into the role of the mid-quarter festivals, and of the corresponding observational solar astronomy, in pre-Christian society.

### The context of solar rituals

Of the solar feasts we have identified, the feast of All Saints is the best example of a feast consciously imposed in opposition to a pagan festival as part of a program of conversion. In this regard, it follows Pope Gregory the Great's proposal of 601 to Archbishop Mellitus of Canterbury:

> not tell us which of the several feasts of the saint is intended, yet his description of the steamy weather combining the night of summer with the dawn of autumn suggests the turning of the seasons, which in Celtic terms would be the beginning of August.

55.  *Vita prolixior S. Justi*, pp. 375–376; cf. 3 Kings 17:4–6. Audin, "L'Omphalos de Lugdunum"; H. Leclercq, "Lyon," *Dictionnaire d'Archéologie Chrétienne et de Liturgie*, vol. 10 (Paris: Letouzey et Ané, 1931), here col. 6.

56.  The cult of St. Justus may have lunar as well as solar elements. Twenty-nine days after the feast of the *adventus corporis sancti Justi* on 4 August is the feast of the *natale sancti Justi* on 2 September.

57.  Although this evidence indicates that an observational solar calendar continued in use, it does not tell us what kind of solar observations were used. Historical and ethnographic analogies suggest that simple horizon observations would have been most likely, although we cannot rule out other techniques.

The temples of the idols among [the English] should on no account be destroyed, but the temples themselves are to be aspersed with holy water, altars set up in them, and relics deposited there. . . . And since they have a custom of sacrificing many oxen to demons, let some other solemnity be substituted in its place, such as a day of Dedication or the Festivals of the holy martyrs whose relics are enshrined there. On such occasions they might well construct shelters of boughs for themselves around the churches that were once temples, and celebrate the solemnity with devout feasting.[58]

Gregory's letter to Mellitus, and Alcuin's letter to Arno of Salzburg noting that all the saints "can shut heaven to the unbelieving and open it to the faithful," both reflect the perspective of leaders of the universal Church, advising missionary bishops how best to supplant pagan traditions with doctrinally orthodox Christian ones.[59]

Correspondence among missionaries tells us little of the perspective of the local community and its lay and clerical leaders; these communities had economic, social, and political reasons to continue important seasonal festivals. The gathering of people from far and near stimulates trade and prosperity; their shared participation in rituals transforms them from isolated individuals into members of a community; the special roles for lay and clerical elites in the pageantry of ritual set them apart and legitimize their status in the community. We will see these concerns, and how they influenced the tension between change and continuity in the ritual calendar, where the members of the local communities can speak to us. They speak most clearly when Christian feasts were not imposed by foreign missionaries, but were developed by members of local Christian communities. The feasts of St. Brigit of Kildare, St. Oswald of Northumbria, and St. Justus of Lyons are examples of this process.

As we have seen from the elements of pagan myth that survived in her *vitae*, Brigit's cult was built upon pagan, as well as Christian, foundations. Considering the rapid rise of her monastery at Kildare, MacAlister suggested that Brigit may have governed a major pagan cult center there, which she transformed into a Christian monastery.[60] Whatever the truth of this speculation, the *vitae* leave no doubt that Brigit had close ties to the Irish nobility. Cogitosus's mid-seventh-century *Life of Saint Brigit* has a clear political agenda, reflecting the dynastic claim by the Uí Dúnlainge for primacy of the bishop, abbess, and church of Kildare over all Ireland.[61]

58. Bede, *Hist. eccl.*, 1.30, tr. Sherley-Price.
59. Alcuin, *Epistolae* 193, *MGH*, Epp., 4, p. 321. The process of deliberate replacement of pagan festivals by Christian ones was discussed by the early church fathers, who are cited in Bede, *De temporum ratione*, 30.39–47.
60. R. A. S. MacAlister, "Tamair Breg: A Study of the Remains and Traditions of Tara," *Proceedings of the Royal Irish Academy*, 34C(1919):340–341.
61. Ó Briain, "Brigitana"; Sharpe, *"Vitae S. Brigitae"*; McCone, "Brigit in the Seventh Century"; Hughes, *Church in Early Irish Society*, pp. 83–88, 113; Cogitosus, *Sanctae Brigidae vita*.

Cogitosus described the assembly at Kildare on Brigit's feast, an assembly reminiscent of gatherings for pre-Christian calendric rituals. At the time of Brigit's feast, Kildare grew from a small monastic settlement to what could be called, considering the numbers of people drawn from throughout the land of the Irish, an exceedingly great city. Cogitosus assured his readers that the power of the saint provided protection within the limits of the city, which she herself had established. Here was total sanctuary for fugitives, and no one need fear an attack by his enemies. Within this city, despite its transitory nature, even the treasures of the king were kept secure.[62]

The assembly on Brigit's feast had all the ingredients of a great religous assembly; there were crowds, cures, an abundance of food, and many gifts and offerings. During this interval of sacred time, a period doubly defined by the death of the saint and by the traditional solar calendar, Kildare had moved outside the world of everyday activities to become a sacred place, protected by the presence and power of the saint.[63] But more was at work here than the power of the saint. Immunities and protection granted by the saint and her successors, the abbesses of Kildare – and implicitly defended by the Uí Dúnlainge kings – contributed to trade and social intercourse, to the profit of the lay and religious communities of Kildare, and to the political prestige of the Uí Dúnlainge.[64]

Association with Christian saints was not essential to the persistence of assemblies such as that held at Kildare; Irish celebrations of the festival of Lughnasa often endured without any close association with Christianity. The greatest assembly, presided over by claimants to the kingship of all Ireland, continued from pre-Christian times at Tailtiu (modern Teltown) in County Meath. It has no strong Christian associations, although a seventh-century account places St. Patrick at the assembly two centuries earlier. Viking raids caused the suspension of the assembly in 873, an event described in the Annals of Ulster as unheard of (*quod non audivimus ab antiquis*).[65] Cúán ua Lothcháin, writing in 1007 of the renewed Lughnasa assembly at Tailtiu, describes its games, its immunities, and its ban on disputes and fixes it in the Julian calendar around the Kalends (i.e., the first) of August.[66] An eleventh-century account of another Lughnasa assembly, at Carman, sponsored by the king of Leinster, adds an economic element to the activities:

> Three busy markets on the ground,
> a market of food, a market of live stock,

---

62. Cogitosus, *Sanctae Brigidae vita*, 39.
63. Hagiographers commonly invoked the power of the saints as protectors of sacred sites and preservers of public peace. Bitel, *Isle of the Saints*, pp. 58–64, 150–152.
64. McCone, "Brigit in the Seventh Century," pp. 108–111.
65. M. MacNeill, *Lughnasa*, pp. 68, 324–338.
66. *Metrical Dindshenchas*, vol. 4, pp. 150–153. Around the year 1000 the solar date of Lughnasa fell on either 31 July or 1 August by the equal-time model, on 2 or 3 August by the equal-angle model.

> the great market of Greek strangers
> where there is gold and fine raiment.[67]

These great assemblies lapsed from history by the end of the twelfth century, yet the Tailtiu festival continued as a local peasant gathering until about 1770.

The theme of royal sponsorship that we have seen in the cult of St. Brigit and the Irish Lughnasa assemblies takes on more direct form in the cult of St. Oswald, king of Northumbria. Oswald's brother and successor, Oswiu, was the moving force behind the establishment of a dynastic cult centered on the martyred king.

After Oswiu transferred St. Oswald's right hand to the Church of St. Peter in the royal city and mercantile center of Bamburgh, the saint's feast came to be celebrated there. We can see in this a continuation of the pre-Christian tradition of political and ritual assemblies at the nearby royal estate of Yeavering.[68] In 1332 King Edward III confirmed Bamburgh's right to a three-day fair, which had been held "from time immemorial" on the feast of St. Oswald.[69] Although we have no documentation of special privileges concerning the fair at Bamburgh, the organizers of English fairs, like their Irish counterparts, typically granted a special "peace" to those attending such assemblies and imposed punishment on peace-breakers. The nearby fair at Durham, for example, was protected by the ninth-century "law and custom of Saint Cuthbert" which extended his peace over seven days travel to and from his fair.[70]

If the saints extended protection to those going to assemblies at their feasts, not even St. Oswald could protect his dynasty and royal city from the declining political fortunes of Northumbria. Early in the eleventh century Oswald's hand and arm were stolen from Bamburgh by a monk of Peterborough, who bemoaned the fact that the decline of the city of Bamburgh from its former royal status had reduced Oswald's feast there to insignificance.[71]

Perhaps the most intricate interplay of solar ritual and cultural change centered on the other saint's feast tied to Lughnasa, the feast of St. Justus of Lyons. St. Justus's feast emerged in a city dedicated to the god Lugh and with a long-established annual ritual in his honor. Soon after the founding of the city the Romans had recognized this local tradition. In 12 B.C. they established a council of all Gaul at Lugdunum, thus continuing this great assembly under Roman aus-

67.   M. MacNeill, *Lughnasa*, pp. 339–344; *Metrical Dindsenchas*, vol. 3, pp. 2–25, 469–480.
68.   The time of year of these pre-Christian assemblies is not known, although Bede's account of Paulinus's visit to Yeavering rules out springtime. Bede, *Hist. eccl.*, 2.14; Brian Hope-Taylor, *Yeavering: An Anglo-British Centre of Early Northumbria*, Department of the Environment Archaeological Reports, 7 (London: Her Majesty's Stationery Office, 1977), pp. 278–281.
69.   *Calendar of Charter Rolls*, 6 Edward III, March 16, 1332 (London: HMSO, 1912).
70.   Edmund Carter, "The Peace of St. Cuthbert," *Journal of Ecclesiastical History*, 8(1957):93–95.
71.   Reginald of Coldingham, *Vita S. Oswaldi*, 2.48.

Figure 13. Altar of Rome and Augustus at Lyons. Reprinted from H. Bazin, *Vienne et Lyon gallo-Romains*, Paris: Hachette, 1891.

pices. On the first of August of that year a great altar (Fig. 13) was dedicated at the confluence of the rivers as one of the first major steps in establishing the new imperial cult of Rome and Augustus. The assembly thus blended the old traditon in honor of the god Lugh with a new one honoring the deified Augustus.[72]

72. Fishwick sees the imperial cult as a Roman response to Hellenistic stimuli and questions the extent of Celtic influence, noting the lack of direct evidence that the earlier Celtic assemblies had been held at Lugdunum on 1 August. Audin sees Celtic influences in the date, the place, and even the nature of the festival. Duncan Fishwick, *The Imperial Cult in the Latin West: Studies in the Ruler Cult of the Western Provinces of the Roman Empire*, vol. 1, pts. 1 and 2 (Leiden: E. J. Brill, 1987), pp. 97–137, 308–350; Audin, "L'Omphalos de Lugdunum"; N. Chadwick, *The*

The festival was sometimes graced by the *adventus* of the emperor himself, a ceremony equated in almost messianic terms with the rising of the Sun as the *oriens augusti*. The emperor came to shed the light of his rule, to display his power, to accept the reverence of his people, and to answer the people's petitions. The annual assembly at Lyons persisted, although undoubtedly cleansed of overt pagan ritual, through the rise of Christianity until the Germanic invasions. As a youth Sidonius Apollinaris, scion of an old Gallo-Roman family who became in turn prefect of the city of Rome and bishop of Clermont, attended on the Consul Asturius and the other great men at the assembly of 449.[73]

The survival of the festival at Lugdunum in its new guise does not mean that the scholarly and priestly community of druids maintained their former status; rather, it reflects the emergence of a new Gallo-Roman elite.[74] In turn, with the decline of both druidism and the empire, the festival would assume new Christian garb with the cult of the saints assuming many functions of the imperial cult. The new Christian leaders of Lyons would replace the imperial *adventus* with the *adventus* of the body of a saint who accepted the reverence of the people, displayed both spiritual and temporal power, and answered the people's petitions.[75]

The earliest record of Christianity at Lyons marks a sharp conflict with the local imperial cult. The annual assembly of all Gaul for the year 177 had been marked by the public torture and execution of a group of Christians for refusing to abjure their faith; commemoration of this event became a central element in the ritual life of the Christian community at Lyons. A cemeterial basilica at Lyons was dedicated to the Machabees, Jewish martyrs executed for their steadfast adherence to their faith and opposition to paganism, including a refusal to sacrifice for the king's birthday.[76] The Christians' celebration of the feast of the Machabees on the first of August provided a clear, if oblique, challenge to the cult of Rome and Augustus.

*Druids*, pp. 71–76; cf. Robert Turcan, "l'Autel de Rome et d'Auguste 'Ad Confluentem,' " *Aufstieg und Niedergang des römischen Welt*, 2, 12, 1 (1982):607–644.

73. MacCormack, *Art and Ceremony in Late Antiquity*, pp. 17, 20–21, 35–37, 45–50, pl. 13; Kantorowicz, "Oriens Augusti"; Strabo, *Geogr.*, 4.3.2; Suetonius, *Claudius*, 2.1; Suetonius, *Caligula*, 20; Dio Cassius, *Hist. Rom.*, 54.32.1; Sidonius Apollinaris, *Epistulae*, 8.6.5.

74. The public display of Celtic calendars at Coligny and elsewhere, on the model of the similar public calendars of Greece and Rome, suggests the waning power of these native observers of nature as their expert control over the ritual calendar was supplanted by a more regular calendric system. Compare the similar circumstances surrounding the rise of the Julian calendar. Bergmann, "römische Kalender."

75. Jacques Fontaine, "Vienne, carrefour du paganisme et du Christianisme dans la Gaule du IVᵉ siecle," *Bulletin de la Société des Amis de Vienne*, 67(1971):17–36, esp. pp. 31–32; Van Dam, *Leadership and Community*, pp. 59–60; Brown, *Cult of the Saints*, pp. 97–100.

76. Brigitte Beaujard et al., *Topographie Chrétienne des cités de la Gaule*, vol. 4, *Province ecclesiastique de Lyon* (Lugdunensis prima) (Paris: de Boccard, 1986), pp. 27–28; Jean-François Reynaud, *Lyon aux premiers temps chrétiennes: Basiliques et nécropoles* (Paris: Imprimerie Nationale, 1986), pp. 21–24, 118–119.

The fourth of August marked the *adventus* of the body of St. Justus to his city; a month later his remains were buried in the cemeterial basilica dedicated to the Machabees. In time St. Justus came to supplant the Machabees as patron of the basilica from which he oversaw his city, and his feast became a major festival at Lyons.[77] The emergence of the cult of St. Justus reinforced this challenge to Roman paganism in a way that epitomized the fifth-century emergence of many local saints in Gaul, as sources of local power replaced the increasingly ineffective and distant Roman rule.[78]

Justus's burial at the season sacred to the abandoned Celtic god Lugh, to distant Rome and the impotent Augustus, and to the alien Machabees marked the advent of a new indigenous spiritual power, drawing crowds in tears and joy to that remote part of the city. Sidonius Apollinaris describes a gathering before dawn of men and women of all classes on the feast of St. Justus. The congregation overflowed the crowded church into the surrounding hilltop cemetery. But this, too, was not just a religious assembly. During the interval between the pre-dawn procession and the mid-morning mass, Sidonius gathered with the leading citizens to engage in the games and small talk that bound together the Christian aristocracy of Lyons.[79] The special commemoration of the *adventus* of St. Justus continued as an important community ritual into the ninth century, when Ado of Vienne mentioned the celebration at Lyons in his martyrology.[80]

Lughnasa provides the best example of the continuation of gatherings, both large and small, at specific ritual centers and during the same festival season over more than a millennium. At most places the continuation of the feast was accompanied by a simple transformation from Celtic to Christian belief. Yet in Lyons the shift was more complex. The Celtic cult of Lugh was supplanted first by the imperial cult of Rome and Augustus and ultimately by the Christian cult of the saints. This cultic change reflected centuries of social change. An indigenous Gallic leadership was supplanted by, or perhaps coopted into, a Roman municipal aristocracy which, in time, became Christian and whose members, as bishops, formed a new Christian aristocracy.

The durability of feasts marking the division of the solar year into quarters displays the continuity of that kind of astronomical practice concerned with the regulation of a ritual calendar that was first revealed in analyses of megalithic stone monuments of the second and third millennia before Christ and left distinct traces in the Calendar of Coligny. In all these cases we see that elements of a prehistoric solar calendar survived into the early Middle Ages, and feasts based on this solar

---

77. *Martyrologe d'Adon*, 248–249, 296. The twenty-nine days from the *adventus* of St. Justus (4 Aug.) to the heavenly birth of the saint (2 Sept.) may have been intended as a response to the dedication of the month of August to the deified Augustus.

78. Van Dam, *Leadership and Community*, pp. 165–172.

79. Sidonius Apollinaris, *Epistulae*, 5.17.

80. *Martyrologe d'Adon*, pp. 248–249, 296.

calendar contributed to the vitality of the ritual centers where they were celebrated.

Those three local saints whose cults arose soon after their deaths, Abbess Brigit
of Kildare, King Oswald of Northumbria, and Bishop Justus of Lyons, display
striking similarities. Their biographers connected all three with attributes of Celtic
deities, and their sanctity included distinct solar elements. They were not ascetics
noted for their piety, but powerful leaders of their communities. As such, their
cults were sponsored by local elites and represented local, rather than universal,
centers of temporal and spiritual power; assemblies for their feasts on the traditional solar mid-quarter days animated and strengthened those communities. The
functions of their local cult centers correspond with the social and economic
functions of astronomically aligned megalithic monuments suggested by conventional archaeological investigations.[81]

The one exception to this pattern, the feast of All Saints, presents a complementary face of the coherence of cult and authority. As a feast of all the saints,
it had no ties to the relics of a particular saint at a local cult center. The ideology
of the cult emphasized this universality and a direct opposition to the pagan solar
cult, rather than the continuation of local, calendrically significant, rituals. Con-
sonant with the cult's theme of universality, its patrons were powerful agents of
imperial and papal authority and advocates of uniformity in ritual, rather than
leaders of local communities.

None of these feasts conform to the romantic notion of a farmers' calendar
based on a folk astronomy that bubbled up from the lower strata of society.
Instead, the ritually significant quartering of the solar year, which shows its earliest
traces in the alignments of prehistoric megalithic structures, continued to be exploited by influential figures from the time of Sidonius Apollinaris to Alcuin of
York.

81.  Burl, "Stone Circles of Cumbria," pp. 183–186, 201–203. This shared pattern, in turn, suggests
     further archaeological investigation of prehistoric calendric sites for evidences of elites, of ritual
     activity, of trade, of influence over extensive communities, and of ties or competition between
     sites with similar astronomical alignments (and presumably similar cults).

# CHAPTER FIVE

# *Computing the central time – the date of Easter*

The whole of nature, which till this moment had the semblance of death, celebrates the Resurrection together with her Lord. The enchanting loveliness of the trees, as they put forth their leaves and are set about, as with gems, by their blossoms. . . . Sol, the kindling of all the stars, lifts up his face and lets it shine, and, like a king in his glory, sets on his head the diadem of the stars. . . . Luna, who sets herself farther away from her rising each day, decks herself for Easter with her full raiment of shining light.

Ps.-Augustine, *sermo* 164, *in Pascha*, vi[1]

Easter is the central event of the Christian drama of salvation, in which human-kind is restored to something like the primordial state at the moment of Creation. Christians have always seen the redemptive act of the Resurrection, like the equinox of spring, as a triumph of light over darkness.[2] According to one tradition, the Passion occurred on the vernal equinox, which was the same day on which Christ was conceived, the same day on which the luminaries were created.

The supernatural mystery of the Resurrection of Christ and the natural mystery of the rebirth of nature combine to place the Easter season at the center of sacred time. Springtime is the season of rebirth, in which the cycle of time returns to the moment of Creation – a return through which nature is re-created once more. In Genesis, on the first day God "divided light from darkness and called the light day and the darkness night" (1:4–5); on the fourth day "He created the two great luminaries . . . to rule the day . . . and the night" (1:16). By tradition the division was equal, with day and night of equal length, and the Moon was created illuminating the whole night. Thus Creation began at the equinox, and the Moon was created full.[3] The Easter ritual must return the worshipper to that central

---

1. Ps.-Augustine, *Sermo in Pascha*, 164.2. I have benefited from the English translation of Brian Battershaw in Rahner, *Greek Myths and Christian Mystery*, pp. 111–112, but I restored the original astronomical metaphors.

2. Anscar J. Chupungco, *The Cosmic Elements of Christian Passover*, Studia Anselmiana, 72 (Rome: Editrice Anselmiana, 1977), pp. 26–54.

3. Ps.-Cyprian, *De pascha computus*, 3.7–10, 4.1; ps.-Augustine, *Questiones veteris et novi testamenti*, 106.4. In a less common interpretation the Sun was created on the fourth day at the equinox, Martin of Braga, *De pascha*, 5; "dicunt enim ideo plenam factam, quia non docebat, ut deus inperfectum aliquid illo dei faceret in sideribus." Augustine, *De Genesi ad litteram*, 2.15.

time. The long debate over the date of Easter pivots around those times of sal-vation and Creation.[4]

Certain themes, which we can usefully arrange under a few headings, contin-ually recur in the course of this debate. Appeals to tradition, to the teachings of their ancestors, and to the practices of their youth are made by learned expositors of Christianity, not unlike the claims of tribal elders discussing traditional ritual practices.[5]

Most important, however, is the theme of uniformity, couched in several dif-ferent ways. Sometimes uniformity was merely a matter of convenience, aimed at avoiding an embarrassing inconsistency.[6] At other times greater issues were seen to be at stake, the harmony of Christendom and the spiritual imperative that all Christians be united in prayer.[7]

With the unity of Christendom we come to an issue with political as well as spiritual implications. The determination of the single uniform day requires the acceptance of some competent authority, either an accepted procedure based on scriptural or astronomical principles, or the decision of senior church officials. Appeals to the authority of popes past and present,[8] of the patriarchs of Alexandria (an authority derived from Alexandrian spiritual preeminence and astronomical

---

4. For historical treatments of the Easter question in the Latin West, see the introduction to C. W. Jones, *Bedae opera de temporibus* (Cambridge: Mediaeval Academy of America, 1943), and the essays in Stevens, *Cycles of Time*. A valuable bibliography on the Easter question is in Strobel, *Texte zur Geschichte des Osterkalendars*.

5. Columbanus, *Epistula 3 ad Papam*, in *Opera*, p. 24.16–21. "That you would grant to us pilgrims in our travail the godly consolation of your judgement, thus confirming . . . the tradition of our predecessors, so that by your approval we may in our pilgrimage maintain the rite of Easter as we have received it from generations gone before."

    Blind adherence to tradition was rejected by Cummian, *De controversia paschali*, 114–119, p. 75. "Our elders, however, whom you hold as a cloak for your rejection, kept simply and faithfully . . . that which they knew to be best in their day, and prescribed for their posterity thus, according to the Apostle [1 Thess. 5:21–22]: 'Test everything, hold fast to what is good, abstain from every form of evil.' "

    Compare from a different culture the Hopi Powamu chief, Intiwa: "Many, many days this has been the custom and we have no right to forsake the ways of our fathers." McCluskey, "Historical Archaeoastronomy," p. 47.

6. Bede, *Hist. eccl.*, 3.25 (tr. Sherley-Price). "It is said that the confusion in those days was such that Easter was sometimes kept twice in one year, so that when the King had ended Lent and was keeping Easter, the Queen and her attendants were still fasting and keeping Palm Sunday."

7. Council of Arles (A.D. 314), *Concilia Galliae, A. 314–A. 506*, p. 5. "Since he 'died and was raised once for all' [2 Cor. 5:14], this time should be observed by all with devout mind. Nor in so great an obedience, may divisions or dissensions in devotion arise. We maintain, therefore, that the Lord's *Pasch* be observed on one day throughout the whole world."

8. Wilfrid, later bishop of York, defended the Roman Easter at the synod of Whitby (A.D. 664). "And even if your Columba – or, may I say, ours also if he was the servant of Christ – was a Saint potent in miracles, can he take precedence before the most blessed Prince of the Apostles, to whom the Lord said: 'Thou art Peter, and upon this rock I will build my Church . . . ' [Matt. 16:18]." Bede, *Hist. eccl.*, 3.25.

expertise),[9] and of church councils[10] all appear in discussions of this issue. Besides such institutional seats of authority, early church teachings were also claimed as authoritative, as were elements culled from the New Testament accounts of the Passion and Old Testament teachings regarding Passover.

Passover is the week-long commemoration of the emancipation of the Jews from their bondage in Egypt. It begins when a lamb that had been selected and set aside for sacrifice on the tenth day of the Hebrew lunar month, Nisan, "the first month of the year" (Exodus 12:2) is slaughtered and consumed on the evening of the full Moon, between the fourteenth and fifteenth days of the month. This Paschal meal begins a week in which leavened bread cannot be eaten, extending from the evening of the fourteenth to the evening of the twenty-first (Exod. 12:1–20; cf. Num. 9:2–3). The time of Passover is further defined as the "first spring month" at Deuteronomy 16:1.

The synoptic Gospels record that the day before the Crucifixion was the first day of unleavened bread, on which Jesus celebrated the Passover meal with his apostles (Matt. 26:17–20; Mark 14:12–17; Luke 22:7–14). This would place the Last Supper on the evening between the fourteenth and fifteenth. In contrast, John's account (18:28, 19:14) seizes on the powerful symbolism of Christ as the sacrificial Paschal lamb to place the Crucifixion itself on the first day of Passover. Although this became, and remains, a matter of dispute, a consensus emerged within the early church that the Crucifixion took place on 14 Nisan.[11]

The Easter practices of the early church varied. In some communities, especially those of Jewish Christians, Easter was celebrated with a Paschal meal on the evening between 14 and 15 Nisan. Elsewhere the feast became detached from the Hebrew luni-solar calendar and was celebrated on the evening of the equinox (25 Mar.) or, to conform with Exodus 12:6, fourteen days later, on 6 or 7 April. Even as late as the fifth and sixth centuries, in northern Gaul a feast for the vigil of the Resurrection was celebrated on the fixed date of 27 March.[12]

Since the gospel accounts record that Christ rose from the grave on the Sunday following the Paschal meal, the practice spread from Rome of celebrating Easter on the Sunday after 14 Nisan. This clearly distinguished the Christian Easter from

---

9. *Prologus Cyrilli*, 2; Bede, *De temporum ratione*, 16.36–38, 16.53–4, 30.49–51, 38.40–42, 42.34–39.
10. Cummian, *De controversia paschali*, 93–4, 98–100. "Added to these was the Synod of Nicea of 318 bishops, who decreed. . . . Again in the Synod of Arles of 600 [*sic*] bishops, which 'first of all' confirmed that, 'concerning the observation of Easter, we should keep it on one day and at one time throughout the entire world.' "
    Bede, *De temporum ratione*, 43. 51–55. "Paternae etenim auctoritatis subsidio fulcimur dum nicaenae synodi scita sectamur, quae quartas decimas festi paschalis lunas tam firma stabilitate praefixit ut decemnovenalis eorum nusquam vacillare, numquam fallere possit."
11. Strobel, *Geschichte des frühchristlichen Osterkalendars*, p. 61 and passim.
12. Strobel, *Geschichte des frühchristlichen Osterkalendars*, pp. 357–374; Gregory of Tours, *Hist. Franc.*, 10.31; Luce Pietri, *La ville de Tours du IVᵉ au VIᵉ Siècle: Naissance d'une Cité Chrétienne*, Collection de l'École Française de Rome, 69 (Rome, 1983), pp. 451–453.

the Jewish Passover, while establishing a Christian claim over their pagan neighbors that Christ, the true Sun of Justice, had risen and shone forth.[13]

The details of these accounts, and their spiritual symbolism, reappear in discussions of the proper date to celebrate Easter. It may be tempting to dismiss these doctrinal disputes either as obstacles to an understanding of the meat of the issue, the astronomy and mathematics of computus, or as mere justifications of cycles chosen for other reasons,[14] but we seek the place of astronomy in medieval culture. The central issue for medieval computists was not to find just *a time* for a celebration, but to reestablish the time of Creation, the time of salvation in which humankind is renewed, to be once again at that time, *in illo tempore*. The astronomy and mathematics served a sacred purpose; the debates over Easter reflect the importance of that purpose.[15]

## Astronomical principles of Easter cycles

As Christianity became separated from its Jewish roots and as a concern with uniformity of belief and practice emerged, leaders of the early church sought to express the date of Easter in terms of the widely accepted civil calendar. We can gain a useful perspective on the historical development of Easter computus by first considering the astronomical and calendric issues one would face in trying to relate dates of the equinox, the subsequent Paschal Full Moon, and Easter Sunday.[16]

Since this problem is concerned not with finding where the Sun and Moon are but with determining the dates of the equinox and the full Moon, we can ignore the places of the Sun and Moon in a geometric model. We only need to consider the average duration of the solar year and the lunar month.[17]

---

13.  Jones, *BOT*, p. 10; Rahner, *Greek Myths and Christian Mystery*, pp. 107–112; Jerome, *In die dominica paschae*, p. 418.7–19.

14.  Neugebauer, *Ethiopic Astronomy and Computus*, pp. 28, 96–98; Jones, *BOT*, p. 338.

15.  Ó Cróinín's description of the Easter controversy as "essentially doctrinal" corrects a common tendancy to dismiss religious concerns. Ó Cróinín, "Pelagianism in Ireland," p. 506.

16.  Throughout late antiquity and the early Middle Ages this problem never drew on any of the advanced astronomical systems that could deal with the inequalities of the Moon's motion and the technical problem of computing the visibility of the new Moon. Ecclesiastical computus remains to this day a simplified analysis of the average motion of the Moon and the average length of a lunar month. Other recent treatments of this technical problem can be found in Harrison, "Luni-Solar cycles"; McCarthy, "Easter Principles."

17.  Ignoring the position of the luminaries almost inevitably leads to the approximations that result from using average periods, since the deviations from the average are most simply related to the position of the Sun and Moon. Both Babylonian arithmetical astronomy and Greek geometrical astronomy, which predicted the positions of the planets or of significant events, dealt with deviations from mean motion. Neither computus nor Maya astronomy, which arithmetically predicted succesive times of events, could account for such deviations.

A general luni-solar cycle is based on the simple observation that after a given number of years, a lunar month will begin on the same date as in some previous year. The interval between these two occurrences provides a luni-solar cycle consisting of a whole number of years and a whole number of lunar months. A very good cycle in which the phases of the Moon recur at approximately the same date in the year is 19 years (or 235 months); other, less precise luni-solar cycles are 8 and 11 years. Two further, nonastronomical, cycles are the cycle of the seven days of the week and the Julian cycle of a leap year every four years. These combine to form the so-called solar cycle of 28 years, after which any given date recurs on the same day of the week.[18] Further relation of the Moon and the calendar requires only a few simple calculations.

For calendric purposes we have to relate the Julian year of 365 1/4 days to a lunar year of 12 lunar months.[19] Since a lunar month is approximately 29 1/2 days long, 12 such months define a lunar year of 354 days. To keep the lunar months at the same place in the solar calendar, we need to periodically intercalate an additional lunar month of 30 days, which produces an embolismic year of 384 days.

It is not difficult to compute when the extra month should be intercalated to make up for the 11-day difference between the Julian year and the lunar year. Thus if the Moon is new on a given date, say 1 January, on the same date one year later it will be 11 days older, next year 22 days older, and the year after that 33 days older, and the intercalary month should be inserted.

A simple and flexible method to compute this change is based upon the epact, that is, the age of the Moon on a specific date. At the beginning of a cycle the epact would be arbitrarily set at zero, or *nullus* as the Latin number system, which lacked a zero, would put it. Each year the value of the epact was increased by 11 days; if the sum would exceed 30, an intercalary month of 30 days was inserted, and the epact was not increased by 11 but reduced by 19, since $11 - 30 = -19$.

Over a full 19-year luni-solar cycle, the epact will have increased by 11 days each year totaling 209 days, while seven intercalary months will have been inserted, reducing the epact by 30 days each time for a total of 210 days. At the end of the cycle the one-day difference is made up by increasing the epact (and the age of the Moon) by one. This *saltus lunae*, or leap of the Moon, brings the epact to *nullus*, the value it had at the first year of the cycle. The month in which the age of the Moon leaps forward in this fashion is thus a day shorter.

---

18. By inspection it can be seen that any date occurs on a given weekday four times in 28 years: after 6, 5, 6, and 11 years. It is only after 28 years that the pattern of weekdays returns in a perfect cycle.

19. We could also use the tropical year of 365.2422 days if we wished to return to the same place in the solar year, e.g., to find repetions of a full Moon on the vernal equinox. Different conventional years would yield different lunar cycles. For example, before the Julian reform the Egyptians employed a lunar cycle of 25 Egyptian years of exactly 365 days each. Neugebauer, *HAMA*, pp. 563–565.

Looked at another way, 19 Julian years equal 6939.75 days. Nineteen lunar years of 354 days equal 6726 days; adding seven intercalary months (210 days) makes 6936 days; the increase in the length of the 4.75 months that include leap days brings the total to 6940.75 days. This one-day excess of the lunar reckoning over 19 Julian years is eliminated by shortening the lunar month including the *saltus*. Although this analysis is expressed in decimal fractions, the computists reckoned the date of Easter using an epact and age of the Moon expressed in integral numbers of days.

While the *saltus* is needed only once for a 19-year cycle; for an 84-year cycle, six such leaps are required; for an 8-year cycle, two would be.[20] This makes these cycles truly cyclic, and to the extent that these cycles accurately represent astronomical reality, the *saltus* should keep the predicted lunar months synchronized with the Moon we see in the sky.

These simple principles of epact, intercalation, and *saltus lunae* guided the cycles used to compute the date of Easter. Before considering the historical development of Easter computus, we should consider how accurately several of these Easter cycles do reconcile lunar months with the Julian and tropical years.[21]

Least adequate of the early Easter cycles shown in Table 2[22] was the 112-year Hippolytan cycle. This cycle has several problems. First, it is unnecessarily long; in its simplest form it repeats itself after only 56 years. Second, since it is based on the 8-year cycle (the *octaeteris* of Greek tradition), which errs by one and a half days in 8 years, it accumulates an error of 21 days over the 112-year cycle.[23]

The other early Easter cycle is an interesting hybrid totaling 84 years. We can consider this as a combination of four 19-year cycles, to which is added an additional 8-year cycle. While the addition of the 8-year cycle detracts from the accuracy of the 19-year cycles, the 84-year cycle is much more accurate than the 112, erring by only 1.28 days. Adjustments to bring an 84-year cycle into agreement with the astronomical full Moon would be required less frequently.[24] Fur-

---

20.  Over an 84-year cycle the epact is increased by $84 \times 11 = 924$ days, while thirty-one intercalary months are inserted, decreasing the epact by 930 days and requiring six insertions of a *saltus*. Over an 8-year cycle the epact increases by 88 days, while three intercalary months are inserted, decreasing the epact by 90 days and requiring two insertions of a *saltus*.

21.  In considering the accuracy of luni-solar cycles, we should avoid the ahistorical assumption that knowledge or use of a given cycle implies *explicit* awareness of the length of the month.

22.  These, and all other computations in this chapter, are based on the ephemerides of the Sun and Moon as given in the *Explanatory Supplement to the Ephemeris*, pp. 98–99, 106–107. The lengths of year and month used by Harrison, "Luni-Solar cycles," differ only in the fourth and fifth decimal places. This insignificant difference arises from his use of P. V. Neugebauer's parameters based on universal time, thereby tacitly incorporating an assumed value for the changing length of a day.

23.  Jones, *BOT*, pp. 11–12; cf. Marcel Richard, "Notes sur le comput des cent-douze ans," *Revue des Études Byzantines*, 24(1966):257–277, esp. p. 267.

24.  The existence of early 84-year tables based on different epacts suggest that occasional empirical

Table 2. *Theoretical accuracy of Easter cycles*

| Cycle | Lunar months | | | Julian years | | | Tropical years | | |
|---|---|---|---|---|---|---|---|---|---|
| Years | Mos. | Days* | Days | Error | Days Century | Days* | Error | Days Century |
| 8 | 99 | 2923.53 | 2922.00 | +1.53 | +19.10 | 2921.94 | +1.59 | +19.87 |
| 11 | 136 | 4016.16 | 4017.75 | −1.59 | −14.45 | 4017.67 | −1.51 | −13.69 |
| 19 | 235 | 6939.69 | 6939.75 | −0.06 | −0.33 | 6939.60 | +0.08 | +0.44 |
| 84 | 1039 | 30682.28 | 30681.00 | +1.28 | +1.52 | 30680.35 | +1.93 | +2.29 |
| 95 | 1175 | 34698.44 | 34698.75 | −0.31 | −0.33 | 34698.02 | +0.42 | +0.44 |
| 112 | 1386 | 40929.39 | 40908.00 | +21.39 | +19.10 | 40907.14 | +22.26 | +19.87 |
| 532 | 6580 | 194311.26 | 194313.00 | −1.74 | −0.33 | 194308.90 | +2.36 | +0.44 |

*Number of days calculated using synodic month of 29.530586 days and tropical year of 365.242285 days, both valid for A.D. 500. During the period A.D. 1–1000 the synodic month was essentially constant, while the tropical year varied by only ±0.000070 days.

thermore, it is a fully recurring cycle after only 84 years, which has obvious advantages when tables computed from it were copied by scribes. Given these advantages, the widespread adoption of various forms of the 84-year cycle should not be surprising.[25]

The more accurate 19-year cycle, which came to dominate medieval computus, appeared in a number of guises. The ideal relationship of the 19-year cycle with the 28-year solar cycle requires a table of 532 years. The labor of computing, or even of copying, such a table is readily apparent. It is not surprising that many early tables based on the 19-year cycle covered only a sequence of five cycles, totaling 95 years.

Besides that of convenience, a 95-year table had a further advantage: it is almost an Easter cycle. Ninety-five Julian years differ by only a quarter day from a whole number of weeks. As a consequence, in three years out of four not only will the

adjustments may have been made to correct discrepancies between the cyclic and astronomical full Moon. O'Connell notes "this error would be noticed all the more readily because of the different results obtained by the Alexandrian [19-year] reckoning." O'Connell, "Easter Cycles," pp. 74–75.

Possibly contributing to exact determination of these errors would be observation of seasonal eclipses. The 84-year cycle, like the 112-year cycle, is related to seasonal eclipses, approximating four and a half revolutions of the lunar nodes. Thus when a seasonal eclipse occurs in the ascending node at the beginning of the cycle, another eclipse will occur in the descending node about the same date, eighty-four years later.

25. The technical variations among the various forms of this cycle are fully treated in Krusch, *84-jährige Ostercyclus*.

Paschal Full Moon fall on the same date after 95 years, it will fall on the same day of the week. Since after 95 years Easter dates repeat in three years out of four, this provides an easy way to check extensions of the tables.[26]

There remains, of course, the question of why Easter cycles were needed at all. Jews got along perfectly well with an observational lunar calendar; what prevented Christians from using such a procedure? One obstacle to direct observation is that the forty-day Lenten fast precedes Easter, and any observations must anticipate Lent sufficiently to allow time to prepare for Ash Wednesday. Furthermore, in sixth-century Gaul the date for the beginning of Lent was announced as early as the feast of the Epiphany, which falls shortly after the winter solstice. Observing the Paschal Full Moon itself to determine the date of Easter would be too late.[27]

Such early announcement, however, could still have been based on earlier observations of the Moon. For example, observations of the age of the Moon at a date near the solstice could directly determine the epact, the central astronomical element of all Easter computus. But when local observations relating the position of the Sun and the age of the Moon have been used to regulate luni-solar calendars in other cultures, they frequently led to disagreements among various observers and communities.[28] Thus it is not an astronomical principle that rules out local observations of the Moon to determine the date of Easter, but the oft-repeated dictum that all Christians everywhere should celebrate Easter on the same day.

### The early history of Easter computus

The definition of the proper date to celebrate Easter is best seen as part of the general drive to define orthodox belief and practice that emerged after the legalization of Christianity by the Edict of Milan (A.D. 313). At the regional Council of Arles (A.D. 314) the bishops of Gaul expressed their opinion that Easter should be celebrated on the same day throughout the world.[29] In his call for the first Ecumenical Council, held at Nicea in A.D. 325, Emperor Constantine asked

---

26.  While 95 Julian years total 34698.75 days, 4957 weeks total 34699 days. On the one instance in four that this 95-year cycle fails, Easter and Passover next recur on the same dates after 12888 weeks (90216 days), approximately 247 Julian years (90216.75 days).

  The 95-year cycle was known in the Ethiopic tradition as the "great Epact cycle" and Neugebauer suggests it was used by computists to check their tables. Neugebauer, *Ethiopic Astronomy and Computus*, pp. 88, 223–225.

27.  Canons of the Council of Orleans (A.D. 541); Canons of the Synod of Auxerre (561 × 605), *Concilia Galliae, A. 511–A. 695*, pp. 132, 265.

28.  Turton and Ruggles. "Agreeing to Disagree"; Malotki, *Hopi Time*, pp. 368–369; Zeilik, "Ethnoastronomy II: Moon Watching."

29.  Council of Arles (A.D. 314), *Concilia Galliae, A. 314–A. 506*, p. 5.

"What could be more beautiful . . . than that this feast day, from which we receive the hope of immortality, be observed by all according to one and the same order and certain rule?"[30]

No specific method of determining the date of Easter can be traced to the Nicene fathers. Their concern seems to have been to separate the Christian Easter from the Jewish Passover on 14 Nisan, while prohibiting the celebration on a fixed solar date as well. Thus they defined, in general terms, the orthodox practice of celebrating Easter on the Sunday following 14 Nisan. This involved a number of astronomical questions. Nisan, the first month, marked the beginning of springtime, but precisely where should it fall in the solar year? Theophilus of Alexandria (A.D. 385–412) stated the accepted view that the Paschal Full Moon should be the first full Moon after the equinox.[31]

It is generally conceded that the Council adopted the then current value of the equinox used in the Alexandrian Easter computus, 21 March (25 Phamenoth in the Alexandrian civil calendar).[32] This differed from the traditional Roman equinox, 25 March, a date that would continue to be recorded in Christian liturgical calendars as the feast of the Annunciation, on which Christ was conceived. There is no evidence to support the assertions of later writers that the Council also adopted a nineteen-year luni-solar cycle.[33] The Council's decision did, however, make the technical problem of determining that single, uniform date more important.

Whereas in the East the practice of Alexandria, based upon a 19-year cycle, became dominant, Rome came to settle on the 84-year cycle as the basis for its Easter computus. Besides an 84-year cycle, Rome continued the traditional equinox of 25 March. Of the various forms of the Roman 84-year cycle that spread throughout Western Christendom during the fourth and fifth centuries, the one of most significance is the Irish cycle with limits for Easter Sunday (the Paschal Term) set between the fourteenth and twentieth of the Hebrew month Nisan.[34]

The discrepancy between Eastern and Western practice came to a head with the Easter of A.D. 455. The Eastern computations called for Easter on 24 April, while the Roman Paschal annals called for Easter a week earlier, on the seventeenth. Pope Leo was well aware of this problem. He noted the discrepancy as early as A.D. 451 in routine correspondence with Bishop Paschasinus of Lily-

---

30.  Eusebius, *De vita Constant.*, 3.18, *PG* 20, col. 1073–76.

31.  Theophilus of Alexandria, *Ad Theodosium Augustum*, 2, p. 223; Bede, *De temporum ratione*, 6.20–25. For a time Rome maintained an alternative view that Easter, not Passover, must fall after the equinox, allowing the Paschal Full Moon to precede the equinox. Jones, *BOT*, pp. 26–27.

32.  Neugebauer, *Ethiopic Astronomy and Computus*, pp. 204–205.

33.  Jones, *BOT*, pp. 17–26.

34.  The Irish 84-year computus maintained an Easter following the equinox on 25 March. McCarthy, "Easter Principles," pp. 218–220. This equinox was defended in ps.-Anatolius, *De ratione paschali*, 2. Bede, *Epistola ad Wicthendum*, 11, in *BOT*, noted inconsistencies in the text and concluded that Anatolius himself had actually defended an equinox on 22 March.

baeum, with whom he had previously discussed the Easter of A.D. 444. After several years of correspondence, Pope Leo conceded that, setting aside all scruples, Easter should be celebrated on the twenty-fourth in the interest of unity and peace.[35]

Clearly, if unity of ritual was important, a permanent solution was in order. Leo's archdeacon, Hilarius, wrote to Victorius of Aquitaine requesting that he prepare a new set of Easter tables. Although Victorius apparently lacked the data required to evaluate the astronomical accuracy of the 84-, 95-, and 112-year cycles, he did recognize that they were mutually inconsistent. Of these three cycles, only the 95-year cycle was consistent with the 19-year cycle used by the Alexandrians, which he held to be more true. Nonetheless, he did not totally accept Alexandrian practice, retaining the Roman equinox of 25 March. He went on to note that this cycle completed its revolution with 532 years, after which the date of Easter repeated itself.[36] The publication (A.D. 457) of his tables covering the period A.D. 28 to 559 introduced into Latin Christendom the first widespread tables based on the 19-year cycle.[37]

While Victorius presented a solution to this element of the controversy, he failed to address adequately the remaining issues that separated Rome from Alexandria. Some unresolved questions concerned only symbolic elements, such as the dates of the Crucifixion, the birth of Christ, and the Creation, and hence the place of those crucial events in the Easter cycle. Others directly influenced the computed date of Easter, among these the exact relationship between the equinox and the full Moon of Nisan and whether the Paschal Term was 15–21 or 16–22 Nisan.

Victorius was aware only of the latter disagreement; in those years where the different Easter limits led to different dates, his tables gave both possible dates for Easter. Otherwise he gave a single date for Easter that generally agreed with the Eastern date. Reinforced by a prefatory letter from Archdeacon Hilarius, the Victorian tables spread, although not universally, in the West. Eighty-four years after their publication the provincial Council of Orleans (A.D. 541) required their use, reserving the resolution of ambiguities to the metropolitan bishops.[38] Yet Victorius's tables did not totally take hold; Gregory of Tours records an anomalous

---

35.  Jones, BOT, pp. 55–61; the correspondence is in Krusch, *84jährige Ostercyclus*, pp. 245–265.

36.  *Prologus Victorii Aquitani*, pp. 18–19, 25.

37.  The fundamental studies of the origin and dissemination of the 19-year tables are Jones, BOT, pp. 61–75; Jones, "Victorian and Dionysiac Paschal Tables." On their dissemination, see Paul Grosjean, "Recherches sur les debuts de la controverse Pascale chez les Celtes," *Annalecta Bollandiana*, 64(1946):200–243, and the studies of Wesley Stevens and Dáibhí Ó Cróinín.

38.  Canons of the Council of Orleans (A.D. 541), *Concilia Galliae, A. 511–A. 695*, p. 132.
     To Harrison, the 84-year interval between the publication of the Victorian tables and their adoption in Gaul "suggests very strongly" a comparison with the older reckoning over an entire Easter cycle. He does not, however, propose any astronomical, scriptural, or historical criteria for such a comparison. Harrison, "Easter cycles in Ireland," p. 318.

Easter date for the year 577 in Spain.[39] Around the year 600 the Irish monk Columbanus repeatedly expressed his preference, and that of his compatriots, for the 84-year cycle over that of Victorius, a tradition which continued in parts of Ireland until 715.[40]

In the meantime (A.D. 525) Dionysius Exiguus, a monk praised by Cassiodorus for his achievements as translator, theologian, and canonist, had prepared a new Easter table which consistently followed Alexandrian practice.[41] Dionysius incorporated the last 19-year luni-solar cycle (A.D. 513–531) from the 95-year table of Cyril of Alexandria, to which he appended an additional 95-year "cycle," extending the tables to A.D. 626. He cautioned his readers that while the tables seemed to repeat after 95 years, this was not a true Easter cycle. Easter did not repeat on the same date after 95 years when the first year was a leap year.[42]

Before Dionysius's tables expired, another writer extended them for a further 95 years to A.D. 721. The Dionysian tables were competently done, but they represented no great mathematical innovation.[43] Their real significance lies in having successfully introduced the Alexandrian Easter reckoning with its 19-year cycle, its equinox of 21 March, and its Paschal Term of 15–21 Nisan into the West. They arrived, moreover, at a significant time; the Victorian tables gave only one date for the Easter of 526, a date that fell on 22 Nisan in violation of Alexandrian principles. That year Dionysius's date was followed at Rome, although, as we have seen, it was some time before his tables completely supplanted those of Victorius, and various Easter tables based on the 19-year and 84-year cycles continued to be used.

## The Irish Paschal controversy

The disagreement over the use of different Easter tables came to a head with Columbanus's conflict with the bishops of Gaul. Columbanus was among the first

---

39.  Ó Cróinín, "Pelagianism in Ireland," p. 514; Gregory of Tours, *Hist. Franc.*, 5.17. For the peculiarity of Gregory's reported Easter date, see Jones, "Victorian and Dionysiac Paschal Tables," p. 412.

40.  Columbanus, *Epistulae*, in *Opera*; Walsh and Ó Cróinín, *Cummian's Letter*, pp. 20–21; Grosjean, "Controverse Pascale chez les Celtes."

    An Irish 84-year Easter table for the years A.D. 438–521 has recently been identified in an early tenth-century manuscript from northern Italy. McCarthy and Ó Cróinín, "Irish 84-year Easter Table." The unusual principles underlying this Irish 84-year table have been fully analyzed in McCarthy, "Easter Principles."

41.  Cassiodorus, *Institutiones*, 1.23.

42.  Dionysius Exiguus, *Praefatio ad Petronio De ratione pascha*, PL 67, col. 487–88.

43.  Neugebauer dismisses Dionysius's tables as requiring no more than an hour's work, transposing dates from the Alexandrian to the Roman calendar. He seems to ignore the possibility that Dionysius might have computed the dates himself, following Alexandrian principles. Neugebauer, *Ethiopic Astronomy and Computus*, pp. 104–105.

of that stream of Irish monks who brought a new form of monasticism to the continent. Ireland's contacts with Rome had been limited; the Irish had never been under Roman political dominion. Their society remained tribal with a fragmented government of petty kingdoms and an intact native scholarly tradition. There is no reason to suppose that the Roman civil calendar supplanted the Celtic solar calendar before the coming of Christianity.

Continental Christianity had developed in the cities; its dioceses followed Roman political subdivisions; and its bishops came from Roman aristocratic families and often considered a bishopric as the culmination of a successful civil career. Irish Christianity, while retaining its doctrinal ties to Rome, was formed by the structure of Irish society. The Celtic holy man, the man of learning, found his Christian analogue in the monk; Irish Christianity took on a distinctive monastic flavor. Since the monastery was primary, the bishop was often limited to a purely sacramental role within the monastery, whose governance was in the hands of its abbot. As the holy man and his lands were tied to the tribe, so did the abbot have ties of kinship to tribal leaders and play a role in dynastic politics.[44]

Around 590, Columbanus and his followers emigrated from Bangor in the north of Ireland to the continent. Here they established the first continental Irish monastery, at Luxeuil, near the border of the Burgundian and Frankish kingdoms. This, and later, Irish foundations became closely tied to Germanic noble families rather than to the bishops of the Gallo-Roman aristocracy.[45] In this context, the celebration of Easter according to the Irish 84-year cycle was not only a spiritual challenge to the unity of Christendom but also a political challenge to the Gallo-Roman episcopate. It could not go unanswered.

Only Columbanus's letters survive to document the controversy. The first is a letter (599 × 600) to Pope Gregory the Great attacking the bishops for following the erroneous table of Victorius; for celebrating Easter before the equinox, the anniversary of the Crucifixion, which he held to be 25 March; and for celebrating Easter on 21 and 22 Nisan, when the Moon rises after midnight and darkness has dominion over night. Columbanus noted that Victorius's tables had been tried and rejected in Ireland by "mathematicians most skilled in the calculation of the computus." He also turned to ancient authority, quoting a treatise on Easter falsely attributed to Bishop Anatolius of Laodicaea that: "It surely is impossible that at Easter any part of darkness should rule over the light, since the feast of the Lord's Resurrection is light, and there is no communion of light with darkness."[46]

As a final flourish Columbanus attacked certain unnamed bishops as Simoniacs, accepting payment for the ordination of priests. A few years later (A.D. 603) he

---

44.  Hughes, *Church in Early Irish Society*, pp. 6–9, 62–64, 76–78; Bitel, *Isle of the Saints*, pp. 85–114, 145–172.

45.  The fundamental study is Prinz, *Frühes Mönchtum im Frankenreich*, pp. 121–151, maps V, VII A, VII B. More accessible is Geary, *Before France and Germany*, pp. 169–178.

46.  Columbanus, *Epist.* 1.3–4; cf. ps.-Anatolius, *De ratione paschali*, 4.

addressed a similar, if more temperate, letter on the Easter question to the bishops of Gaul, then meeting in council at Chalon-sur-Saône. He summarized his earlier arguments on Easter and added the telling critique that the Victorian tables fail to resolve the issue when it is most important, that is, when dates are in doubt.[47] The Council's decision is not recorded, but Columbanus soon wrote again to Rome, requesting that he and his followers, pilgrims "dwelling in seclusion, harming no one," be permitted to follow the practices of their predecessors.[48]

Columbanus's letters reflect, and by bringing them to the attention of Rome may have made more significant, the divergence between insular and continental Easter dates. With the rest of Western Christendom, the Irish and British had adopted the earlier 84-year Easter cycle; they were not ready to abandon it without careful consideration. Victorius's tables had been known, and rejected, in Ireland by the last decade of the sixth century. About A.D. 629 Pope Honorius I wrote to the Irish, urging these "few, located at the farthest ends of the earth" not to celebrate Easter "contrary to the paschal computus and synodal decrees of the whole world's bishops."[49]

In response the Irish bishops held a synod at Mag Léne (ca. A.D. 630), which had almost reached a decision to "celebrate Easter with the universal Church" when a speaker, whom Cummian described as a " 'whited wall' pretending to 'preserve the tradition of our elders,' " disturbed the consensus. The synod dispatched a delegation to Rome, "the chief of cities," to resolve the issue. The Irish delegation found their Easter reckoning to be inconsistent with the date celebrated at Rome by Greeks, Hebrews, Scythians, and Egyptians. Upon the return of the delegation, the monk Cummian wrote a letter summarizing the events and defending a 532-year cycle against the recalcitrants in the north of Ireland who maintained the 84-year cycle.[50]

Less than a decade later the issue arose again, this time through a papal letter to the Irish bishops spelling out the theological implications of unorthodox Easter dates. The northern Irish celebrated Easter on 14 Nisan, the day of the Crucifixion, while the Resurrection was the sign of redemption. To celebrate Easter before the Resurrection was to imply that grace was unnecessary for salvation, recalling the ancient heresy of Pelagianism.[51]

It is in this context that Irish scholars directed to the date of Easter that in-

---

47.  Columbanus, *Epist.*, 2.5.
48.  Columbanus, *Epist.*, 3.2.
49.  Bede, *Hist. eccl.*, 2.19.
50.  Cummian, *De controversia paschali*, 250–288. On the identity of Cummian and those to whom he wrote, see Walsh and Ó Cróinín, *Cummian's Letter*, pp. 7–15.

     Cummian's well-chosen phrases from Scripture emphasize the limits of following tradition. The term "whited wall" had been hurled by St. Paul at the high priest Ananias, that God should strike him down for misuse of the law (Acts 23:2–3). Jesus had accused the scribes and Pharisees of hypocrisy in preserving the traditions of their elders (Mark 7:1–6).
51.  Bede, *Hist. eccl.*, 2.19; Ó Cróinín, "Pelagianism in Ireland."

novative curiosity they applied to the languages and knowledge of the Mediterranean world. The Irish did not fear pagan Roman learning, and Christianity had not disrupted Irish scholarly traditions; thus Irish scholarship blended elements from Roman and Irish traditions. In astronomy this blending led to the practical art of computus, which would become a central element in a new pedagogical framework.

Given its numerical methods and its focus on calculating the date of the Paschal Full Moon and the subsequent Easter Sunday, the Irish used terms for number and calculation to describe this art. The term that survived in Latin was *computus*, but Irish writers claimed that specifically suited to the problem of the Paschal computus were two words derived from Irish roots: *rima*, number, and *rimarius*, calculator. It seems that even before they were faced with the Christian problem of Easter, the Irish had a name for this kind of calendric calculation.[52]

To clarify the Easter question, in the middle of the seventh century Irish scholars assembled a collection of continental and Irish texts bearing on the issue. These texts are, with the exception of an excerpt from Macrobius's *Saturnalia*, entirely Christian in origin.[53] Their astronomical concerns are focused more on the religious symbolism of the Sun and Moon than on matters of astronomical observation or theory. This collection, long associated with the Irish Easter controversy, has recently been traced to a group of scholars working in the region near Kildare in southeast Ireland.[54]

This scholarly community showed other interests in relating astronomy and ritual. From it comes the naturalistic exegesis found in the Irish Augustine's *De mirabilibus sacrae scripturae*, whose author notes that Joshua's miraculous interruption of the Sun's motion must also have stopped the Moon, so as not to disrupt the Easter cycle, which always returns on itself after 532 years.[55] This community is also the source of Cogitosus's roughly contemporary *Vita* of St. Brigit of Kildare, which draws on elements from the prologue to Victorius of Aquitaine's discussion of Easter, while championing Brigit's feast and the monastic *paruchia* at Kildare. We have already seen the association of an earlier pagan tradition of solar

---

52.   *De ratione conputandi*, 3.7–13. See also Walsh and Ó Cróinín, *Cummian's Letter*, p. 33, n. 126, pp. 117–118, n. 3.8.

  *Rima* appears to be purely Celtic in origin. Nils Törnqvist, *Zur Geschichte des Wortes Reim, K. Humanistika Vetenskapssamfundit i. Lund Årsberättelse*, 1934–35, p. 9. Prof. James Marchand kindly resolved a number of my concerns concerning the origins of the root *rim-*. E-mail interchange on MEDTEXTL, 25–26 July 1991.

53.   Jones, "Sirmond Manuscript"; Jones, *BOT*, pp. 105–113.

54.   Ó Cróinín, "Irish Provenance of Bede's Computus," pp. 234–238. The location is established by the appearance of the name of a prominent Leinsterman, Suibine mac Commáin, in a dating clause in the manuscript. Smyth would make the collection later, placing the ambiguous dating clause between 658 and 669, which then forms a *terminus post quem* for the collection. Marina Smyth, "Isidore of Seville and Early Irish Cosmography," *Cambridge Medieval Celtic Studies*, 14(1987):69–102, here pp. 95–96.

55.   *De mirabilibus sacrae scripturae*, 2.4, PL 35, col. 2175–76.

ritual with her feast. Cogitosus, who has been suggested as the author of *De mirabilibus*, may be a major figure in this astronomical learning.[56]

The same scholarly milieu also gave rise to a series of introductions to the principles of computus.[57] But where Martianus Capella and Boethius had treated astronomy as a mathematical discipline closely related to geometry, within the framework of the seven liberal arts, here computus was discussed as a specialized practical art in a different framework. One of these texts, the *De ratione conputandi*, lists four arts as required within the church: Sacred scripture, history, numbers (i.e., computus), and poetics. These four arts add Christian scriptures to the genealogical lore, natural knowledge, and poetry taught orally by the *fili*, the druid, and the bard, with all now directed towards Christian ends.[58]

To fit this new framework, these texts introduce new sources into their discussions of the Paschal question. Boethius's *De arithmetica* is quoted on the nature of number; Isidore of Seville's *Etymologiae* and *De natura rerum* on astronomy and other topics; Macrobius's *Saturnalia* on the structure of the calendar; and the usual patristic sources on the date of Easter. These new materials expanded the study of computus so that finding the date of Easter became the guiding rationale for an introductory course in astronomy.

Topics heretofore taken for granted in discussions of Easter now entered the discussion. Whereas the church fathers had addressed Roman audiences, for whom the calendar was part of everyday experience and the rudiments of astronomy part of their background in the liberal arts, the Irish had to present the elements of arithmetic, the divisions of time, and the basics of astronomy to students who found them all new and foreign. In presenting these elements they concentrated upon those that, while not necessarily of immediate practical value, would help the student understand the question of Easter. The Sun goes through a sign of the zodiac in thirty days, ten hours, and one or two *puncti*, and the Moon in two days and eleven hours; the Moon is closer to the earth than is the Sun, hence the Moon runs its course more rapidly.[59]

In some of their examples we see hints of the traditional Celtic solar year, already indicated where pagan festivals were transformed into Christian feasts. The

---

56. Cogitosus's authorship is affirmed by Ó Briain, noting that MacGinty denies it in his unpublished edition, which I have not seen. Ó Cróinín merely notes the similar use of sources in Cogitosus's *Brigidae vita* and the *De mirabilibus*. Ó Briain, "Brigitana," pp. 133–134, Ó Cróinín, "Irish Provenance of Bede's Computus," p. 240.

57. Most of these texts remain unedited; my discussion is based on Walsh and Ó Cróinín's citations from their work on the the *De ratione conputandi*.

58. *De ratione conputandi*, 2.8–11. Columbanus had earlier recommended four similar types of learning in his letter to Sethus in Columbanus, *Opera*, p. 186: "Sint tibi divinae dogmata legis [scripture] / Sanctorumque patrum casta moderamina vitae [history] / Omnia quae dociles scripserunt ante magistri [?] / Vel quae doctiloqui cecinerunt carmina vates [poetics]."

    Poetry frequently intrudes into Irish-influenced classifications of the liberal arts. Díaz y Díaz, "Arts Libéraux Espagnols et Insulaires," pp. 43–44.

59. *De ratione conputandi*, 55, 68.2–4, following Isidore of Seville, *De natura rerum*, 4.1, 19.1.

*De ratione conputandi* spends almost as much time on the four seasons of the year, noting that they are divided midway between the solstices and equinoxes, as on the various kinds of solar and lunar months.[60] In contrast to Isidore's definitions of the civil year (*annus civilis*) as either a period of twelve months when a star returns or a solar year of 365 days, here the civil year is presented as a solar year regulated by "signs" at the four corners of a city or on its walls that mark the courses of the Sun and Moon and the solstices and equinoxes. Alternatively, the civil year is regulated by the four directions (*quatuor partes celi*).[61]

A continued concern with horizon observations is indicated by the technical names assigned in the *De ratione conputandi* to the relative positions of the setting Sun and new Moon as they change through the course of the seasons. The Sun was found below the new Moon (*subincensio*) at the short days near the winter solstice, above the new Moon (*supraincensio*) at the long days near the summer solstice, and at a middling position (*mediaincensio*) near the equinoxes.[62]

### The Paschal controversy in England

By the middle of the seventh century the students flocking to the verdant scholarly pastures of Ireland were as numerous as the twinkling stars gathering around the pole of heaven. Irish proficiency in ecclesiastical computus and a wide range of other subjects was attracting students from England and Gaul.[63] These students and the Irish monasteries on the Celtic fringe of Britain provided ready avenues by which Irish computistical learning reached England, where a similar controversy over Easter had arisen. Perhaps it was best expressed in the Northumbrian court, at the frontier between the Roman practice and that shared by the northern Irish and Britons. Queen Eanfled and her companions followed the Roman practice she had learned in Kent, whereas King Oswiu, who had been educated by Irish monks, followed the Celtic practice with the rest of his court. Since the

---

60. *De ratione conputandi*, 47–51; on the seasons beginning between the solstices and equinoxes he quotes the Irish pseudo-Anatolius, *De ratione paschali*, 14.

61. *De ratione conputandi*, 46.26–34; the unpublished seventh-century Irish computistical collection, the *De ratione temporum*, 27, as quoted in Walsh and Ó Cróinín, *Cummian's Letter*, p. 156, n. 26; cf. Isidore of Seville, *De natura rerum*, 6.2, 6.4.

   The most likely interpretation of these as directional markers for observation is reinforced by the texts' alternative of regulating the *annus civilis* by the celestial four directions. A less likely interpretation would be of some form of *parapegma*, although there is no mention of stars and there is no practical reason for placing public calendars on all four walls of the city rather than in a central location. Ó Cróinín could not identify any classical source for these means of regulating the civil year; they seem characteristically Irish.

62. *De ratione conputandi*, 65. Bede expresses a similar concern with the Sun and Moon on the horizon. Bede, *De temporibus liber*, 10.12–18; *De temporum ratione*, 38.40–48; letter of Ceolfrith of Jarrow to Nechtan, in Bede, *Hist. eccl.* 5.21.

63. Bede, *Hist. eccl.*, 3.27; Aldhelm, *Epistulae* 5.

Celtic Easter was sometimes earlier and sometimes later than the Roman, in one year the queen continued to fast while the king had finished his fast and was celebrating Easter.[64]

To resolve this disagreement, King Oswiu held a synod in A.D. 664 at the monastery of Whitby, governed by Abbess Hilda. After comparing the Irish and Roman positions, the English, like the southern Irish at Mag Léne, adopted the Roman Easter. Oswiu based his decision on the religious arguments of the unity of Christendom and the power of Peter and his successors as keepers of the keys of heaven, rather than on any question of astronomical precision. Nonetheless, the decision was not universally followed, and its full acceptance waited on the detailed astronomical justification provided by the Northumbrian monk Bede of Jarrow (A.D. 672/3–735).

The adjacent monasteries of Jarrow and Wearmouth followed a rule based on that of St. Benedict, rather than an Irish rule. Their founding abbot, Benedict Biscop, traveled repeatedly to Rome and brought back books on a wide range of subjects. In addition, the monasteries had a copy of the collection of computistical texts compiled by the Irish to teach the issues surrounding the Easter question. It is in this center of learning that Bede, who entered the monastery as a young boy of seven years, received his education.[65]

Despite the isolation of his monastic life, Bede corresponded actively with his colleagues, some of whom investigated the archives in Rome to further Bede's historical research. Bede's scholarly career is characterized briefly by Alcuin of York:

> This famous scholar wrote many works,
> unravelling the mysterious volumes of Holy Scripture,
> and composed a handbook on the art of metre.
> He also wrote with marvelous clarity a book on time,
> containing the courses, places, times, and laws of the stars.
> He was the author in lucid prose of books on history,
> and the composer of many poems in metrical style.[66]

---

64. Bede, *Hist. eccl.*, 3.25. Although a social inconvenience and an example of disunity, the story of the two fasts suggests something more. The Irish had a pre-Christian tradition of fasting, not as an expression of asceticism, but as a means of enforcing a claim against a superior. Such a fast was not just a hunger strike but also a magical act against one's rival. It could be answered only by a counter-fast, in which the contender who first gave up the fast was defeated. This practice continued into the Christian era. Such fasts were directed against pagan kings, and even against God, by the early Irish saints. D. A. Binchy, "A Pre-Christian Survival in Medieval Irish Hagiography," pp. 165–178 in Whitelock, McKitterick, and Dumville, *Ireland in Early Medieval Europe*.

That this story records an instance when the queen's fast continued after her husband's had ended suggests an intent to show the victory of the Roman party. Whatever the pious Christian intent of Eanfled's fast, it could easily be seen by her contemporaries as a magical victory.

65. Bede, *Hist. eccl.*, 5.24.

66. Alcuin, *Saints of York*, 1306–12; see also the list of his works in Bede, *Hist. eccl.*, 5.24.

Bede's works recapitulate the traditional categories of learning discussed by the Irish computists: Scripture, history, numbers, and poetics. His successors lavished praise upon his works. Notker the Stammerer of St. Gall, advising a student about commentators on the Song of Songs, gave Bede those celestial attributes usually granted Christ. "God, founder of the order of nature, who on the fourth day of the world's Creation brought forth the Sun from the east, on this sixth age of the world has placed [Bede as] a new Sun in the west to illuminate the whole earth."[67] The quality and scope of Bede's scholarship epitomized the new light of learning shining from England to the continent.

Bede wrote five or six works on the Easter question, justifying the Dionysian tables in terms of both the old and the new learning. There is a distinct difference in style between his computistical treatises, designed to teach computus, and his letters, which present arguments for the Roman Easter. In his letters Bede discussed theological matters, citing Scriptures, the church fathers, and those early computistical writers who wrote in a similar vein. In his computistical treatises Bede added the natural learning of the continent to the older computistical tradition, expanding the bounds of computus to treat astronomy in even greater detail.[68]

Chief among the fonts from which Bede drew astronomical learning was Pliny's *Natural History*,[69] supplemented by Virgil's *Georgics* and those sources already known to the Irish: Macrobius's *Saturnalia* and Isidore's encyclopedic works. Yet he also drew secular learning from such religious sources as Augustine's commentary on the Psalms and Ambrose's *Hexaemeron*.

Bede's main debt to these sources concerns the rudiments of Greek cosmology: the planets, the zodiac, the causes of lunar and solar eclipses, and the zones on the earth defined in terms of the ratio between longest and shortest days. Since the equinoxes are central to the date of Easter, Bede related the changing position of the Sun in the zodiac to the solstices, the equinoxes, and the changes of the seasons. In this he followed Pliny in placing the Sun in the eighth degree, rather than at the beginning, of the appropriate zodiacal sign at the equinoxes and solstices. The existence of inconsistent locations of the solstices and equinoxes would pose a challenge to later computists. Bede attributed to Hippocrates and unnamed pagan and Christian wise men the Roman norm that the solstices and equinoxes fall on 25 December, 25 March, 24 June, and 24 September, noting that those are the dates of Christmas, the Annunciation, and the birth and conception of

---

67. Notker Balbulus, *De interpretibus divinarum scripturarum, PL* 131, col. 996.
68. His computistical works include *De temporibus* (A.D. 703), a letter to Plegwin (A.D. 708), a letter to Helmwald (n.d.), *De temporum ratione* (A.D. 725), and a letter to Wicthedus (A.D. 725 × 731). Bede may also have drafted Ceolfrid's letter to Nechtan (ca. A.D. 710).
     The first five can be found in Jones, *BOT*; Ceolfrid's letter is in Bede, *Hist. ecd.*, 5.21.
69. Eastwood notes that Bede's direct knowledge of Pliny can be confirmed only for Book Two, Pliny's principal book on astronomy. Eastwood, "Plinian astronomy," p. 201. Borst claims that Bede's scientific and religious writings depended more extensively on Pliny. Borst, *Plinius und seine Leser*, pp. 98–110.

John the Baptist. However, he ultimately accepted the view of "the Egyptians," who hold the palm among calculators, that the vernal equinox falls on 21 March.[70]

Bede then attributed the changing height of the Sun at noon and the changing length of daylight to the sphericity of the earth and the Sun's changing position along the inclined path of the zodiac. These new materials, like those introduced by the Irish, are of little practical value for Easter computations, but they deepen the student's understanding of the astronomical concepts underlying the Easter question. This is not the advanced astronomy of a Ptolemy, but it does introduce into the curriculum of English monastic schools those concepts from which Ptolemy's astronomy had grown.

Like his Irish predecessors, Bede spoke of observations not found in his classical sources. He suggested to his students an empirical means to demonstrate that leap years are needed. If one were to note carefully where the Sun rises in the east (*a medio orientis*) at the vernal equinox, one would find that each successive year it rises somewhat lower (i.e., more to the south of east). If a leap year is inserted after four years, then on the day of the equinox the Sun will rise again at its former place.[71]

A closely related example appears in the letter from Abbot Ceolfrith of Jarrow to Nechtan, king of the Picts, who wished support for his decision to change from the Irish to the Roman practice.[72] This letter presents a simple operational definition of the Paschal Full Moon, noting that the equinox Sun is indicated by its rising due east (*a medio Orientis*), after which the Paschal Full Moon would be

70.  If, following Greek astronomical practice, we fixed the equinox at the beginning of Aries, then at Bede's time when the equinox fell on 16 or 17 March, the Sun would enter the eighth degree of Aries on 24 or 25 March, near the Roman equinox. Given Bede's scanty knowledge of spherical astronomy, he had no means to discern this.

Pliny, *Hist. nat.* 2.17.81; Bede, *De temporum ratione*, 30.1–44, 38.40–42, *De temporibus liber*, 10.12–14; *Epistola ad Wicthendum*, passim.

71.  Bede, *De temporibus liber*, 10.12–18; *De temporum ratione*, 38.35–48.

Bede's proposed observations of change in a single year are doubtful, as they approach the limits of observational precision. The point of equinoctial sunrise at Jarrow changes by about 10' of arc after a single year; the change over four years of 40' would clearly be observable. Ethnographic examples attain a precision in the range of ± 20', occasionally reaching values of ± 10' under ideal conditions. McCluskey, "Historical Archaeoastronomy," p. 42; Zeilik, "Ethnoastronomy I: Calendrical Sun Watching," pp. S19– S21.

72.  The text is given by Bede, *Hist. eccl.* 5.21. The similarity between this unusual discussion of observational criteria for the equinox in the letter and in Bede's authentic works argues strongly for Bede's authorship. The many similarities with Bede's known writings led Plummer to conclude that Bede drafted the letter. C. Plummer, ed., *Baedae Historia ecclesiastica gentis Anglorum* (Oxford, 1896), vol. 2, p. 332.

Jones rejected Plummer's view, attributing the similarities to "common sources and a common computistical jargon." Jones, *BOT*, p. 104, n. 4. Recent commentators have taken the compromise view that Bede probably contributed to the letter or revised it for inclusion in his history. B. Colgrave and R. A. B. Mynors, *Bede's Ecclesiastical History of the English People* (Oxford: Clarendon Press, 1969), p. 534, n. 1; Wallace-Hadrill, *Bede's Ecclesiastical History*, p. 196.

indicated when it rises from due east as the Sun sets. The idea that the Paschal Full Moon can be seen opposite the Sun, and therefore at the place of the Sun at the autumnal equinox, had appeared previously in the Irish computistical literature.[73] The original contribution of this letter is to move from abstract concepts to phenomena that are more easily grasped and may, in principle, be observed: the Moon is full when it rises at the time of sunset; a full Moon is the Paschal Full Moon when it rises, like the equinoctial Sun, from due east.

Besides the place of moonrise, Bede also discussed the appearance of the rising Moon. He explained the changing orientation of the lunar crescent, which Pliny, Isidore, and many others had held to be a weather sign, in terms of the relative orientation of Sun and Moon at different seasons.[74] He explained, more clearly than the Irish, that the Moon and the Sun appear low when they are in southern signs, and high when they are in northern signs, and hence the Moon sometimes appears higher than the Sun although it is actually closer to the earth.[75] Here we see the new concepts of geometric astronomy explaining familiar observational practices.

The examples that Ceolfrith and Bede present reflect the same kind of observations of the Sun and Moon along the horizon that have been indicated in studies of British, Scottish, and Irish megalithic alignments.[76] In particular, Ceolfrith's inclusion of such observations among the arguments that Nechtan could use to convince his subjects suggests a continuity of practice among Nechtan's advisors and followers. Bede and Ceolfrith's sensitivity to this kind of observation indicates that even those who spent most of their lives within the confines of a monastery were familiar with such remnants of pagan astronomical traditions.

Bede's works represent the culmination of the art of computus. Many later writers would expound on his subject and add to it, but Bede had shaped computus into its definitive form. A body of literature that had begun with debates over the theological symbolism of the Resurrection, the changing seasons, and the great luminaries in the heavens had, while retaining its theological focus, been transformed into a general introduction to astronomy and an essential part of early medieval learning. His works spread rapidly on the continent, brought with them by Irish and Anglo-Saxon scholars. By the ninth century, as the reckoning of the date of Easter became a part of the drive for clerical education and liturgical uniformity sponsored by Charlemagne and his successors, an educated cleric would be expected to know the arithmetical techniques and astronomical foundations of computus.

---

73. Ps.-Anatolius, *De ratione paschali*, 2, p. 319.
74. Bede, *De temporum ratione*, 25; advocates of the common belief are cited in Jones, *BOT*, p. 360.
75. Bede, *De temporum ratione*, 26; cf. *De ratione conputandi*, 65.
76. Despite frequent indications of lunar observation, the archaeological surveys yield few signs that the Sun or the Moon were observed at equinoctial rising. Aubrey Burl, "Chambered Tombs"; Ruggles et al., *Megalithic Astronomy*; Ann Lynch, "Astronomical Alignment or Megalithic Muddle?" pp. 23–27 in D. Ó Corráin, ed., *Irish Antiquity: Essays and Studies Presented to Professor M. J. O'Kelly* (Cork: Tower Books, 1981).

# CHAPTER SIX

# Observing the celestial order – monastic timekeeping

At the night office during one season, winter, the nocturns are chanted before cockcrow, as the prophet says: "at midnight I rose to give you praise" [Ps. 118:62]. And of summer he likewise says, "At night my spirit keeps vigil for you, O God" [Is. 26:9]. . . . What is imperative in winter is that cockcrow follow after the nocturns are finished, because the nights are long. . . . In spring and summer however, that is, from Easter to 24 September, which is the winter equinox, the brothers are to begin nocturns at cockcrow because of the shortness of the nights.

*Rule of the Master* (A.D. 520 × 530)[1]

Monasteries, when considered merely as social institutions with stable rules of governance and possessing in their libraries the remnants of ancient learning, were a principal source of stability and the chief focus of learning during the unsettled period from the fourth to the tenth century. Yet monastic culture was primarily a religious culture dedicated to knowledge and worship of God. Within monastic communities the principal use of secular learning, including knowledge of the heavens, was to serve those communities' multifaceted religious needs.[2]

Monasticism, like Christianity itself, was transplanted to the Latin West from the Orient, especially from Egypt. In Egypt monasteries had been established in the desert, in spiritual quarantine from the Greek culture of the cities. In the West sympathetic bishops often sponsored the establishment of monasteries near their cathedral cities. These monasteries in turn provided the Christian formation of many later bishops, some of whom were conscripted from monastic contemplation into the more active service of the episcopacy.

Among the founders of Western monasticism was Martin of Tours (ca. 316–397) who had served in the Roman army before resigning in the year 356 to take up a more rigorous form of service. He lived first as a hermit near Milan, and then he moved to Gaul, where he established a monastery at Ligugé with the support of his mentor, Bishop Hilarius of Poitiers. In 371 he was called by the Christians of Tours to be their bishop. Soon after arriving, he established a mon-

---

1. *Rule of the Master*, 33.1–2, 7, 10.
2. This discussion is based on McCluskey, "Gregory of Tours"; cf. Borst, *Ordering of Time*, esp. chap. 4.

astery at Marmoutier, across the Loire a short distance upstream from the city. Although it housed a monastic community, this refuge was more a collection of separate cells than the integrated complex of buildings that made up later monastic cloisters. In this we can see the continued influence of Eastern monasticism.[3]

St. Martin, as a soldier who had turned to the cloister, provided a suitable patron for the Franks, who under Clovis adopted Catholic Christianity. By the fifth century the influx of Germanic peoples had fundamentally transformed the ancient world. Roman governance in the provinces, long decaying, was replaced by the Germanic kingdoms, of which the most long lasting would be the kingdom of the Franks. Yet some provincial aristocrats continued to carry out their civic responsibilities by finding positions of suitable status within the emerging structure of the Christian church.[4]

While Martin provided a soldier-monk as a model for the Frankish aristocracy, monks took their model from John Cassian, who had lived in the monasteries of Egypt and Palestine during the closing decades of the fourth century. After a short stay in Rome he founded two monasteries, one for men and the other for women, at Marseilles. In the 420s Cassian wrote his *Institutes* at the request of Bishop Castor of Apt; these provided spiritual and practical guidance for the monastery he had founded there and for many later generations of monks.

Italy provided the chief center for Western monastic legislation. Most influential was the *Rule*, which St. Benedict of Nursia (ca. 480–550) drew up for his followers at Monte Cassino. Benedict's rule is not an independent composition; he drew from Cassian's *Institutes*, which he recommended to his monks, and especially from the anonymous *Rule of the Master*, written near Rome earlier in the century.[5]

Running as common threads through these rules are the values of humility, obedience, and a well-regulated life of prayer and work. But monasteries were also influential centers of Christian learning. Several of the monks educated at Lérins, an isolated monastery on a Mediterranean island off the coast of France, left the cloister to become bishops.

Learning also flowed from the secular to the monastic world. Cassiodorus Senator (ca. 490–580), who sought refuge in the monastic life, contributed significantly to the development of monasteries as intellectual centers. Retiring from high government service during the political turmoil associated with the end of

---

3. Stancliffe, *St. Martin and His Hagiographer*, pp. 22–26. Jean Hubert, "L'Érémétisme et l'archéologie," pp. 473–475 in *L'Eremetismo in Occidente nei secoli XI e XII*, Atti della seconda Settimana internazionale de studio, Mendola, 30 Agosto–6 Septembre 1962 (Milan: Vita e pensiero, 1965); repr. in his *Arts et vie sociale de la fin du monde antique au Moyen Age* (Geneva: Librarie Droz, 1977).

4. Prinz, *Frühes Mönchtum im Frankenreich*, pp. 31–33; Geary, *Before France and Germany*, pp. 140–147.

5. On the chronology of and influences among monastic rules, see Adalbert de Vogüé, "The Cenobitic Rules of the West," *Cistercian Studies*, 12(1977):175–183.

the reign of Theodoric the Ostrogoth, Cassiodorus established a monastery at Vivarium, his estate in southern Italy. There he wrote *Introduction to the Divine and Human Readings*, a guide to sacred and secular literature that provided centuries of monastic librarians with a list of essential books. We will consider Cassiodorus's influence on the preservation of scholarly astronomy in the next chapter.

## Monastic prayer and astronomy

Insights into monastic timekeeping are found in the monastic rules which regulated the monks' lives and thereby documented the daily round of communal activities: working together, sharing meals, and especially communal prayer, which occupied much of the monks' day and night. The gathering of the community for these shared activites required some signal to announce the assembly and a means of regulating the time to sound the signal. Monastic communities thus developed a number of timekeeping practices, watching the Sun by day and the stars by night, to regulate the hours of monastic prayer.

This timekeeping, like calendar keeping and computus, is an applied rather than a pure form of astronomy; it is best understood in the context of the rituals it was intended to regulate. The daily course of the Sun served monks, as it does the members of other traditional communities, as the chief guide to the course of daytime activities. But in addition to repeated prayer through the day, monastic rules and Sacred Scripture extended the round of prayer to the middle of the night and before dawn.[6]

Thus monks prayed in the dark of night when the chief luminary was no longer available to indicate the passing hours. They rose before dawn to complete nocturns before the cry of the cock heralding the coming daybreak. Some knowledge of the changing length of day and night, of the changing appearances of the stars with the seasons and the Moon with the month, was essential for planning and regulating the daily round of prayer.

The earliest Western monastic rules recognized that the changes of the seasons called for variation of the regular cycle of work and prayer: relief from work in the long hot summer afternoons, consolidation of nocturnal prayers in the short summer nights, division of prayers by a time of restful contemplation in the long dark nights of winter.[7] The rules' recognition of the changing length of day opened the way for further consideration of astronomical phenomena.

The earliest requirement that the time of prayer be governed by watching the stars is found in John Cassian's discussion of the night offices, largely based on

6.  Psalm 118:164; Psalm 118:62; Isaias 26:9; *Rule of the Master*, 33.1–2, 34.1–3; *Regula Benedicti*, 16.1–5.
7.  *Rule of the Master*, 33.1–14, 27–42; 50.1–71; *Regula Benedicti*, 8–10; 48.2–13.

the practices he had observed in the monasteries of Lower Egypt.[8] Cassian's *Institutes* and *Regula* disseminated this practice throughout the Latin West. Cassian stipulated that the night offices should be governed not by a whim or by the sleeplessness of the monk assigned to wake his brothers, but by careful and frequent observation of the course of designated stars that regulate calling the brothers to prayer.[9]

Subsequent monastic rules, blending Cassian's inspiration with a more legalistic tradition drawn from Roman civil law, shared Cassian's concern with the regular sequence of monastic prayer. The *Rule of the Master* assigns this duty in a weekly rotation among the deans, or *praepositi*, who stay up in pairs lest one fall asleep, waking the abbot when the time for prayer has arrived. He, in turn, sounds the signal to wake the remaining monks. The later *Rule* of St. Benedict holds the abbot personally responsible for sounding the signal to wake the monks for prayer, although he is allowed to entrust this duty to a careful brother. In the event the community is wakened too late to complete their prayers, the negligent monk should "make due satisfaction to God in the oratory."[10] Isidore of Seville's *Rule for Monks* (written in the first quarter of the seventh century) assigns the responsibility for nocturnal timekeeping specifically to the sacristan.[11]

Regular communal prayer had been reemphasized by the bishops of Gaul at the Second Council of Tours, held in 567. They spelled out the number of psalms to be chanted at each season of the year in the Basilica of St. Martin and stipulated that whoever failed to chant the required number of psalms would fast on bread and water until evening prayer.[12] The mechanics of reckoning the time for prayer, however, are not specified in monastic rules or conciliar *canons*; they leave such details to local authorities.

Keeping time by watching the stars is a rudimentary kind of astronomy, more akin to the astronomies of primitive peoples and of Hesiod's *Works and Days* and Virgil's *Georgics* than to the learned mathematical astronomy of the quadrivium for which Ptolemy's *Almagest* would provide the paradigm. Yet it does require

---

8.  Taft, *Liturgy of the Hours*, pp. 58–62, 76–80, 96–100; O. Chadwick, *John Cassian*, pp. 71–73; Cassian, *De institutis coenobiorum*, 2.1–3.6.

    It would be tempting to see the timekeeping practices of Egyptian monasteries as a Christian adaptation of the ancient Egyptian practice of reckoning the hours of the night by observing the rising of specific constellations, the decans. Resolution of Egyptian monastic timekeeping practices would demand investigation of the largely unexplored Coptic texts.

9.  "Sollicite frequenterque stellarum cursu praestitutum congregationis tempus explorans," Cassian, *De institutis coenobiorum*, 2.17; cf. *Regula Cassiani*, 16, p. 175.

10.  *Rule of the Master*, 31–32; *Regula Benedicti*, 46.1, 11.12–13. It may be significant for the spread of stellar timekeeping that Benedict specifically recommends Cassian's *Institutes* as a guide for monastic perfection; *Regula Benedicti*, 73.6.

11.  Isidore of Seville, *Regula monachorum*, 20.1, PL 83, col. 889–890.

12.  *Concilia Galliae, A. 511– A. 695*, pp. 182–183. The stipulated monthly sequence of psalms is not the same as that of *De cursu stellarum*.

familiarity with the constellations and some knowledge of their movements with the seasons and of the changing length of day and night through the year.

The earliest of the few discussions of these technical details appears in Gregory of Tours's short treatise *On the Course of the Stars*. The emphasis upon the proper completion of the daily round of prayer clearly motivates Gregory's application of astronomical knowledge to the service of God. *De cursu stellarum* was written some time after his consecration as bishop in 573, apparently as part of his episcopal obligation to instruct and guide the clergy and regulate the monasteries near Tours.[13] His treatment of astronomical details compensates for their absence from monastic rules.

In *De cursu stellarum* we see a practical astronomy that has been reduced to its barest essentials. Gregory tells us he intended to provide only those elements of astronomy that would help a monk determine the proper time to rise in the hours before dawn for nocturnal prayer. But this astronomy was not indigenous to Celtic or Frankish Gaul. Gregory was an educated member of a Gallo-Roman senatorial family that had provided four generations of bishops to the church. He knew the declining scholarly tradition of Roman Gaul, from which he selected those elements of astronomy that served the practical religious needs of monastic culture and his faith in a divinely established natural order.

## The celestial order

The circles and poles characteristic of the geometric astronomy of the liberal arts do not appear in Gregory's introduction to astronomy. Instead of using abstract geometrical coordinates, Gregory defined the positions of the stars in terms that would be directly useful to an observer watching them rise and set.

If Gregory did not present the geometric details of ancient astronomy, his work benefited from that astronomical tradition. Gregory located the constellations in relation to the path of the Sun, following the pattern of the Sun's movement set forth in Martianus Capella's widely circulated *Marriage of Philology and Mercury*. Gregory had read this standard introduction to the liberal arts and noted that Martianus "taught you . . . to trace the courses of the stars in his astronomy."[14]

13.    Numerous monasteries were founded near Tours before 590, inspired by the establishment of Marmoutier by the Monk-Bishop Martin (ca. 316–397). Prinz, *Frühes Mönchtum im Frankenreich*, map XIIa.

The exceptional influence of bishops on the foundation and regulation of early monasteries, in contrast to the later independence of abbots under the *Rule* of St. Benedict, is noted by Élie Griffe, "Saint Martin et la monachisme gaulois," in *Saint Martin et son temps: Mémorial du XVI^e centenaire des débuts du monachisme en Gaule, 361–1961*, Studia Anselmiana, 46 (Rome: "Orbis Catholicus"/Herder, 1961), pp. 17–19.

14.    Martianus Capella, *De nuptiis*, 872; Gregory of Tours, *Hist. Franc.*, 10.31.

Christian culture guided Gregory's relation to the diverse astronomies of his predecessors and contemporaries. Gregory explicitly mentioned Virgil and the other astronomical poets, if only to reject their mythological descriptions of the constellations.[15] He likewise rejected those constellation names that assign mythological origins to the constellations or would impute a divine nature to the stars. In *De cursu stellarum* the celestial motions do not represent powers independent of God's dominion over creation; they, like the other natural wonders that Gregory discussed, are under the dominion of God.

The divine lawgiver's dominion over nature, and nature's ultimate dependence upon him, arise in monastic prayer as well. The pre-dawn office of lauds takes its name from Psalms 148–150, a series of praises of God by his creation that form one of the earliest unvarying elements of monastic prayer. Through the centuries monks began each day by calling upon the heavens and all creation to praise God for his dominion:

> Let all these praise the Lord; it was his command that created them.
>
> *He has* set them there unageing for ever; *given them a law which cannot be altered*[16]

Medieval commentators commonly interpreted this passage as referring to an unchanging natural order,[17] yet the divine dominion was still questioned among the rustic population of Gregory's time who, like their Germanic and Celtic forebears, continued to address the Sun and the Moon as "lords."[18] Even the eighth-century Italian scribe who made the oldest surviving copy of *De cursu stellarum* was influenced by a syncretistic conflation of Christianity with astral theology. Accompanying the discussions of the regular change of the length of day through the seasons and of the visibility of the Moon through the month are two crude drawings of personifications, or even deifications, of the Sun and the Moon set in circles atop pillars. The image of the Sun contains a bearded figure crowned with twelve solar rays, representing the months or the signs of the zodiac and reflecting the Sun's role as governor of the year (Fig.14). The figure of the Moon is feminine and crowned with a lunar crescent.[19]

Like his contemporaries, Gregory explicitly rejected as a superstitious art, incompatible with Christian culture, those teachings of the mathematical astrologers

15. Gregory of Tours, *De cursu Stellarum*, 17.

16. Psalm 148:5–6; my emphasis. The text is from Ronald Knox's translation of the Vulgate (New York: Sheed & Ward, 1954).

17. Augustine, *Enarrationes in Psalmos*, CIII, Sermo 4, 2; CXLVIII.3; Cassiodorus, *Expositio Psalmorum*, CXLVIII. 1–6; an anonymous Irish commentator, *Hiberno-Latin Gloss on the Psalms*, p. 307; and Remigius of Auxerre, *In Psalmos*, CXLVIII, PL 131, col. 840.

18. Eligius of Noyon, *Sermo*, in Audoenus of Rouen, *Vitae Eligii liber II*, p. 707.7–9.

19. Bamberg Staatsbibliothek, MS Patr. 61, fol. 79r; on the provenance of this manuscript, see Lowe, *Codices Latini Antiquiores*, vol. 8, 1029. On early Christian symbolism of the Sun and Moon, see Rahner, *Greek Myths and Christian Mystery*, pp. 89–176.

Figure 14. Personification of the Sun. Gregory of Tours, *De cursu stellarum*, Bamberg Staatsbibliothek, MS Patr. 61, fol. 79r. Italian, late eighth century. Used by permission.

who sought to foretell the future through the stars.[20] If Gregory rejected an astrology that treated the stars as causes, he nonetheless admitted that there could be signs in the heavens. His discussion of the constellations closes with a brief treatment of comets, where he acknowledged them as signs of future events. As examples he described two recent comets, one of which had preceded a pestilence in the Auvergne while the other had preceded the death of King Sigibert in 575.[21]

In the same passage he quoted Prudentius's poem for the Epiphany, which interpreted the Christmas star as both foretelling the coming of Christ and symbolizing his dominion over the heavens. Gregory followed Prudentius in making this careful distinction: he accepted signs in the heavens, but he would not admit that these signs were independent causes that challenged God's dominion over nature and humankind. Yet Gregory's opposition to astrological causation lacks the strident tone that reflected Prudentius's conflicts with a still active Roman paganism.[22]

The monks for whom Gregory wrote were members of a society that was, at least nominally, Christian. To the extent that the Christian universe operated under the dominion of a supreme lawgiver, its laws were potentially intelligible.

## The astronomy of De cursu stellarum

De cursu stellarum can be divided into four sections, of which only the last two touch our concern with astronomical practice. The first two sections of De cursu treat what seems to be a scarcely related topic, the wonders of the world. Yet here Gregory goes beyond the immediate demands of monastic timekeeping to emphasize God's trustworthy dominion over nature.

The first section deals with the seven wonders, or miracula, of the ancient world and the second with seven natural wonders created by God which will endure "until the Lord directs the dissolution of the world."[23] Gregory treated these natural wonders largely as spiritual allegories; some, such as the burning springs of Grenoble, were presented as instances of the divine hand suspending the usual course of nature.[24] Yet the last two natural wonders, the regular motions of the

---

20.   Gregory of Tours, De cursu stellarum, 16.

21.   Gregory of Tours, De cursu stellarum, 34; cf. Hist. Franc., 4.36(51); 4.24(31). On Gregory's evaluation of the significance of signs and prodigies, see de Nie, "Roses in January."

22.   Prudentius, Cathemerinon, 12.5–40; cf. Apotheosis, 611–630.

23.   Gregory of Tours, De cursu stellarum, 9.

24.   " 'Limpharum in gremiis inimicus condidit ignis, communesque ortus imperat alta manus. . . . ' O admirabile potentiae divinae misterium!" Gregory of Tours, De cursu stellarum, 14.

      The first two lines are a quotation from a poem by a certain Hilarius, usually taken to be Bishop Hilarius of Arles (d. 449), but sometimes Bishop Hilarius of Poitiers (fl. ca. 315–367). In speaking of fire hidden in the bosom of the spring, Hilarius and Gregory allude poetically to the conflict between pagan and Christian attitudes towards nature, for Lympha is not just a term for a spring but the name of a goddess of such places.

Sun and of the Moon, were noted as clear expressions of a divinely established order.[25] These then lead into the two sections on the stars, which first describe the motions of certain constellations and then show how the regular course of the stars can be employed to regulate the regular course of monastic prayer. Despite their differences, these four sections reflect the unifying theme running through the entire treatise: a consideration of how God established the order of creation, and how it, in turn, can order the worship of the Creator.

The astronomical discussion begins with the regular changes of the length of the day through the seasons, reckoning this *miraculum*, along with the changing phases of the Moon, as the sixth and seventh wonders of God's creation. Yet the scheme that Gregory provided for the length of daylight has Mediterranean, not northern European, origins. It reflects a simple method of calculation, in which the length of daylight in December is taken as nine hours and increases regularly by one hour per month until June when it reaches fifteen hours. The same linear process then continues back to December when the length of daylight is again nine hours.

Such rudimentary arithmetical schemes were commonly found in antique culture in both agricultural and religious calendars. Numerous examples of what may be a survival of early Babylonian linear schemes employ the same 15:9 ratio of longest to shortest days. This ratio is more appropriate for the Mediterranean region than for the latitude of Tours. Yet given the simplicity of changing one hour each month, systems employing this ratio are widespread.[26]

Gregory described a similar linear scheme to define the changing duration of moonlight through the month (Table 3). Here the Moon stays up for about one hour more each night until it is visible all night long on the fifteenth day of the month; it then begins to vanish, rising later each night, until it is not seen at all on the thirtieth day of the month. This schematic thirty-day month had been discussed by Vettius Valens and would appear again in Bede's *De temporum ratione* and other computistical texts. This method does not account for seasonal changes, and Gregory made no attempt at precision, rounding his tabulated values, albeit inconsistently, to the next lower half hour.[27]

---

25. Gregory of Tours, *De cursu stellarum*, 15–16. Gregory's acceptance of an accustomed and recognizable order of nature is discussed by de Nie, "Roses in January," pp. 267–279; "The Spring, the Seed, and the Tree."

26. Neugebauer, *HAMA*, pp. 706–711. At the latitude of Tours (47.3° North) the ratio of longest to shortest day is approximately 15.75:8.25. Pliny (*Hist. nat.*, 6.216, 219) describes the longest day of 15 hours as appropriate to northern Greece and southern Italy, while he gives the longest day for the Atlantic coast of Gaul as 16 hours and for Britain, 17 hours.

   Similar linear functions appear in other early medieval Latin texts. An anonymous martyrology in a ninth-century computistical anthology employs a similar linear function stepping one hour twice each month. McCulloh, "*Martyrologium Excarpsatum*," pp. 179–237.

27. Gregory of Tours, *De cursu stellarum*, 18; Neugebauer, *HAMA*, p. 830; Bede, *De temporum ratione*, 24. Quite different, and inconsistent, values are given in Pliny, *Hist. nat.*, 2.11.58; 18.65.323–325.

Table 3. *Lunar visibility (hours)*

| Day | Vis. | Day | Vis. |
|-----|------|-----|------|
|     |      | 30  | 0    |
| 1   | ½    | 29  | ½    |
| 2   | 1½   | 28  | 1½   |
| 3   | 2    | 27  | 2    |
| 4   | 3    | 26  | 3    |
| 5   | 4    | 25  | 4    |
| 6   | 5    | 24  | 5    |
| 7   | 5½   | 23  | 5½   |
| 8   | 6    | 22  | 6    |
| 9   | 7½   | 21  | 7½   |
| 10  | 8    | 20  | 8    |
| 11  | 8½   | 19  | 8½   |
| 12  | 9½   | 18  | 9½   |
| 13  | 10   | 17  | 10   |
| 14  | 11   | 16  | 11   |
| 15  | 12   |     |      |

In discussing the constellations, Gregory was careful to avoid any taint of the astral theology or astrology of his pagan Greek and Roman predecessors. Gregory stated that he would not use the traditional constellation names, and his constellations sometimes overlap the boundaries of several classical constellations. Yet he was careful to describe them clearly so that his reader could identify the proper stars for pre-dawn observations.[28]

Gregory's descriptions of the constellations may include as many as three separate elements, all aimed at making the stars useful for keeping time. The earliest copy, in an Italian manuscript from the late eighth century, continues this practical concern by showing each constellation as an observer would see it rising on the eastern horizon. The drawing of the Greater Cross, which Gregory substituted for the classical constellation Cygnus the Swan, matches Gregory's description of it lying on its side as it rises (Fig. 15), the Cross being raised only at the end of the night in the latter days of the world. Although Gregory clearly saw the constellation as a cross with the constellations Alpha and Omega pendant below its arms, the scribe did not use this obvious artistic motif in his illustrations (Figs. 16 and 17). Artistically, the work of the Bamberg scribe is crude, and there is no

28.   The traditional identifications are those of the astronomer J. F. Galle in Haase's edition of *De cursu stellarum*, pp. 42–49. Alternative identifications have been proposed by Bergmann and Schlosser, "Gregor von Tours und der 'rote Sirius.'" For defenses of the traditional view, see the letters of S. C. McCluskey and R. H. van Gent, "The colour of Sirius in the sixth century," *Nature*, 325(1987):87–89.

Figure 15. The constellation *Crux major*. Gregory of Tours, *De cursu stellarum*, Bamberg Staatsbibliothek, MS Patr. 61, fol. 79v. Italian, late eighth century. Used by permission.

attempt to portray the figure that the stars signify; yet the orientation of the stars and the brightness of distinctive ones are clearly indicated.[29]

For most constellations, the month of its first appearance in the western sky before sunrise is given. For six of them Gregory also tabulated the number of hours that the constellation would be visible for each month through the year, mentioning, when appropriate, those months when the constellation is invisible or those when it appears twice each night, setting after sunset and rising again before dawn. Sometimes when a description does not directly give times of rising Gregory noted that a constellation rises a given number of hours after another constellation in the series.

Some descriptions indicate, in general terms, the constellation's path through the heavens and consequently the place where it rises or sets. In four cases, Gregory noted explicitly that a constellation follows the same path that the Sun does in a particular month; at other times, there are more general allusions to it following a northerly or southerly path.

29. These figures differ from the more artistic, but astronomically unintelligible drawings of the classical constellations in manuscripts of Aratus and Hyginus (see Fig. 19). This striking difference reflects the concern with actual astronomical practice by readers of *De cursu stellarum*.

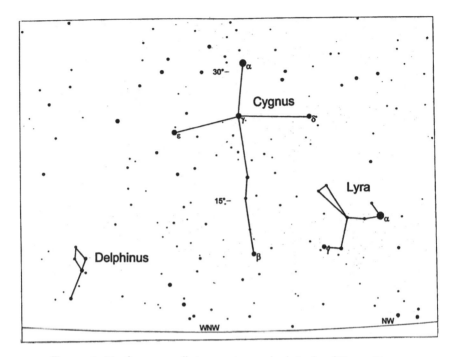

Figure 16. Northern constellations setting at the latitude of Tours. Cygnus corresponds to Gregory of Tours's Greater Cross, Delphinus to Alpha, and Lyra to Omega. Star positions precessed to the epoch of A.D. 575 using SKY-MAP® Software, version 2.1.

Gregory then turned from an abstract discussion of the constellations to show how constellations can be used to regulate the time monks should rise for the first of the two night offices: nocturns (called matins in modern monastic practice), which begins a few hours after midnight, and matins (now called lauds), which ends about dawn. Allowing for the seasonal changes of nighttime and daylight, Gregory gave for each month several rules of the sort that if one begins when a given star rises one can sing a certain number of psalms, or if one begins when another star reaches the third hour, one can complete another number of psalms.[30] Comparison of these rules with the computed appearances of the stars shows that the rules were based on observations made near the latitude of Tours at the time Gregory wrote. Thus Gregory was not merely repeating received astronomical

30.   Gregory of Tours, *De cursu stellarum*, 36–47.

Figure 17. Exaltation of the Cross. Note the letters Alpha and Omega hanging from the arms of the cross and the stellate florets in the background. Eighth century, Musée de Narbonne. Used by permission.

material; his descriptions reflect the application of astronomical observations to the problem of regulating the time of prayer.[31]

Observations of the rising of stars, such as those Gregory prescribed to regulate the time of prayer, are best carried out when the observer has an unobstructed view of a fairly distant horizon. At Gregory's time monks frequently resided in scattered buildings or caves; Jean Hubert maintains that the quadrangular cloister, with its view of the horizon restricted by church, dormitory, refectory, and other buildings, did not become the dominant style of monastic architecture in Gaul before the eighth century. St. Martin's foundation at Marmoutier, a few kilometers from Tours, exemplifies this early practice. Its site on the north bank of the Loire offers the unrestricted view of the horizon that Gregory's astronomy presumes.[32]

## Monastic timekeeping after *De cursu*

Gregory's instructions suggest that implicit in early Western monasticism was an institutional framework for the continued practice of astronomy. This kind of astronomy reflected and reinforced the monastic values from which it sprang: obedience to the *Rule*, which provides a framework of uniform ritual, within the monarchial autonomy of the abbot. The *Rule* and the heavens provide a uniform order, which the abbot or his representative observes to call his fellows to prayer.

The continuing importance of this astronomical tradition is reflected in the commentaries on Benedict's *Rule* emerging out of the Carolingian monastic reform. Early in the ninth century, Smaragdus of St. Mihiel quoted John Cassian's *Regula* on the importance of accurately calling the monks to prayer. Unlike Cassian, Smaragdus did not propose any method of timekeeping. He did mention the existence of an *horologium*, apparently a sundial, when criticizing monks who treat fasting as a punishment and "frequently look at the *horologium* and often lift their eyes to heaven, noting the course of the Sun."[33]

Later in the century, Hildemar of Milan drew on a rudimentary form of astronomical knowledge to resolve an ambiguous passage in Benedict's *Rule* concerning the time of monastic prayer.[34] In winter the *Rule* requires that monks rise at the eighth hour of the night. Hildemar noted that some consider this excessively harsh since in winter the night is eighteen hours long, which seems to require

---

31. McCluskey, "Gregory of Tours," pp. 18–19.
32. Stancliffe, *St. Martin and His Hagiographer*, pp. 22–26. Note, however, that Martin's earlier foundation at Ligugé, near Marseilles, was in the ruins of a former Roman villa and employed the enclosed quadrangular plan typical of Roman architecture. Stancliffe, p. 23.
33. Smaragdus of St. Mihiel, *In Regulam S. Benedicti*, pp. 270, 96–97.
34. Hildemar, *Expositio Regulae*; cf. the shorter South Italian version in *Pauli Warnefridi diaconi Casinensis in Sanctam Regulam Commentarium* (Typis Abbatiae Montis Casini, 1880).

that monks rise before midnight and pray ten hours until dawn. He resolved this problem by noting that at the equinoxes night and day both have twelve hours and that Benedict intended this pattern of twelve hours of night to be continued throughout the year.[35]

Hildemar also commented specifically on the practice of timekeeping. In discussing the passage where Benedict imposed the responsibility for announcing the time for prayer upon the abbot or a responsible brother appointed by him, Hildemar noted that the brother should not be arbitrarily chosen, but should be responsible and know well the hours of the night. Where Benedict had stipulated that a monk who wakens the community late for prayer should "make due satisfaction to God in the oratory," Hildemar took the opportunity to discuss how the severity of the offense can vary with circumstances. The least severe penance would be assigned for a case in which clouds obscured the stars, and the monk, through fear of sounding the signal too early, instead sounded it too late.[36] It is striking that Hildemar did not mention any alternative timekeeping device or technique that this conscientious monk could have consulted.

Peter Damian (ca. 1067) recommended what was probably the more common practice, that when the stars were not visible, the *significator horarum* should chant psalms to note the passage of time. This obligation of keeping time by watching the stars is also seen in rules derived from the Cluniac reform, which specifically assign this duty to the sacristan.[37]

Observation of the stars seems to have been the principal means of nocturnal timekeeping at most ninth-century monasteries. The hour glass was not yet used, although some monasteries did possess water clocks. As early as the tenth century the prosperous monastery of Fleury (St. Benoît-sur-Loire, founded 651) used a water clock (or clepsydra) for timekeeping.[38] This clepsydra was probably a simple one, unlike earlier Greek and Roman models. Abbo of Fleury (ca. 945–1004) discussed, following Macrobius, how to measure the quantity of water flowing from a clepsydra to time the rising of the signs of the zodiac. He described a clepsydra as a bronze vessel with a very small opening near the bottom, with no indication of an attached pointer to measure the flow of time or of the more

---

35. *Regula Benedicti*, 8.1–3; Hildemar, *Expositio Regulae*, pp. 277–279.

36. *Regula Benedicti* 11.11–13, 47.1; Hildemar, *Expositio Regulae*, pp. 288–289, 475.

37. Peter Damian, *De perfectione monachorum* 17, PL 145, col. 315; *Horologium stellare monasticum*, p. 5, n. 8.

 The sacristan's role as timekeeper can be traced back to Isidore of Seville and is shown in England in the tenth-century *Regularis concordia*, 1.27, 2.29, and in Lanfranc, *Monastic Constitutions*, pp. 82–85.

38. R. T. Balmer, "The Operation of Sand Clocks and Their Medieval Development," *Technology and Culture*, 19(1978):615–632, places the earliest evidence of the hour glass at the beginning of the fourteenth century and in a nautical, rather than monastic, context.

 Keeping time with a clepsydra is mentioned in Fleury's tenth-century customary; *Consuetudines Floriacenses antiquiores*, p. 42.

sophisticated analemma used in some ancient clepsydras to account for the varying length of seasonal hours.[39]

Besides using clepsydras for timekeeping, monks used other instruments to refine their observations of the stars. Archdeacon Pacificus of Verona (776–844) described a simple instrument to determine the time of night by watching the rotation of the bright star in the handle of the little dipper (Polaris, or α *ursa minoris*) around a faint fifth-magnitude star that marked the pole at the year 800. This instrument's demonstration of the turning of the stars around the geometric axis of the world, and the use of classical constellation names (*plaustrum minus* and *arctos minor*) in the texts that describe it, bring elements from the astronomy of the liberal arts into the tradition of nocturnal timekeeping.[40]

Even after the turn of the millenium, when the astrolabe first arrived in Latin Christendom, and at least into the thirteenth century, the practice of observing the heavens without instruments continued. An eleventh-century collection of liturgical *cantica* from a French monastery, in a pocket-sized volume apparently intended for a monk's private use, includes descriptions of how to observe the changing azimuth of stars over the buildings of the monastic enclosure to determine the time of nocturnal prayers throughout the year.[41] Fragments of a similar set of instructions, written about 1267 on loose pieces of slate at the Cistercian Abbey of Villers-en-Brabant, describe how the sacristan was to tell time by watching the Sun and stars as they appeared at various windows. At Villers these astronomical observations were supplemented by a water clock, which was set after the evening prayer of compline.[42]

The techniques described here represent a willingness to adapt traditional astronomical practices to the changing circumstances of northern European monasteries. The earlier Greco-Roman technique of watching risings of stars, which Gregory of Tours had recommended, is more precise with nearly vertical rising over an unobstructed local horizon. Now observations were made over nearby

---

39.   Evans and Peden, "Natural Science in Abbo of Fleury." Dr. Peden has graciously provided me with a copy of her transcription of this passage from Berlin, MS Phillipps. 1833, fol. 13v. Cf. Macrobius, *In Somnium Scipionis*, 1.21.11–22.

      On more advanced clepsydras see Vitruvius, *De architectura*, 9.8.8–14. Neugebauer, *HAMA*, pp. 869–870, describes the surviving fragments of two Roman clepsydras from Salzburg and Grand (France). On the subsequent significance of the clepsydra, see North, "Monasticism and Mechanical Clocks."

40.   Wiesenbach, "Pacificus von Verona." Although Pacificus's instrument is depicted in several manuscripts, no examples survive. At the year 800 Pacificus's pole star was within 35' of the pole; by the year 1000 it was still within 1°13' of the pole. During that period Polaris (now the pole star) was separated from the pole by 7° to 8°.

41.   *Horologium stellare monasticum*; Poole, "Monastic Star Time Table."

42.   D'Haenens, "La clepsydre de Villers." The Villers slates indicate how many uniform units of time, as measured by the clepsydra, should pass before rising for psalms on different nights of the year.

structures, employing the more rapid change of azimuth at northern latitudes to compensate for the restrictions on the horizon imposed by the new design of monastic cloisters.

These changes in the practical astronomy of monastic timekeeping reflect an evolving tradition that was open to a variety of locally appropriate observing techniques and instruments. Yet the universal commitment of this tradition remained the observation of the stars so that their orderly course, established by God, would be reflected in an orderly round of prayer, decreed by the *Rule*.

# CHAPTER SEVEN

# Astronomy in the liberal arts

But think how trivial and empty such [earthly] glory is. You know from astronomical computation that the whole circumference of the earth is no more than a pinpoint when contrasted to the heavens; in fact, if the two are compared, the earth may be considered to have no size at all. . . . Do you therefore aspire to spread your fame and significance when you are confined to this insignificant area on a tiny earth? How can glory be great that is severely limited by such narrow boundaries?

Boethius, *On the Consolation of Philosophy*[1]

In late antiquity, scholars maintained that the appreciation of nature and its laws has a valuable place in a liberal education, a conviction that many still share. The most powerful science for the ancients was astronomy, both because of the nobility of the celestial objects that it studied and because of the predictive power by which its geometric models displayed the enduring order of the universe. It should come as no surprise then, that astronomy was the only science incorporated into the antique framework of the seven liberal arts. Seen as one of the liberal arts, astronomy was not concerned with mundane matters of time and the calendar. Rather, astronomy was seen as a part of philosophy that demonstrated the geometrical structure of the celestial realm and, thereby, provided a model from which one could learn to live well.

When considered as a part of philosophy, astronomy stressed the value of the orderly structure of the celestial spheres, while the computational techniques that generations of Greek astronomers had derived from this geometrical model played little role in late antique education. Only rarely do early medieval discussions of astronomy refer to this kind of astronomical calculations. In the sixth century Cassiodorus vaguely described how astronomers perform their calculations, noting that they correct their computations by adding or subtracting a second computation to the first computation of the body's position.[2] The allusion is to the techniques used with Ptolemy's *Almagest* or *Handy Tables*. Yet although Cassiodorus mentioned these calculations, he did not teach his reader how they were done; for such details he directed the interested monk to more authoritative

---

1. Boethius, *De consolatione philosophiae*, 2, pros. 7; transl. Richard Green.
2. Cassiodorus, *Institutiones*, 2.7.2.

works. He cited only two astronomical authorities: the Roman encyclopedist, Varro, whose lost introduction to the liberal arts is briefly mentioned, and Ptolemy, to whom Cassiodorus assigned first place among the Greeks.[3]

A Latin edition of part of Ptolemy's *Handy Tables*, entitled the *Preceptum canonis Ptolomei*, survives in several medieval manuscripts, the earliest of which was copied about the year 1000. The instructions to these tables were edited in the sixth century, as they contain sample calculations for the years 528 and 534. Since the tables were computed for the meridian of Alexandria, using the Egyptian year of 365 days, the instructions included examples for computing the difference in time between Rome and Alexandria. Cassiodorus's description corresponds quite closely with this edition of the *Handy Tables*, which suggests that the Latin translation may have been prepared for the school that he and Pope Agapetus had unsuccessfully tried to establish at Rome.[4] Nonetheless, for the next four centuries the use of such tables for astronomical calculations leaves no trace in the Latin-speaking world.

The computists, who would certainly have been interested in the difference between the Alexandrian year of 365 1/4 days and Ptolemy's Egyptian year of 365 days, show no awareness of this distinction. Both Bede, who certainly ranks as the master of medieval calendric knowledge, and Hrabanus Maurus described the sequence of Egyptian months in terms of the Alexandrian reckoning without mentioning any other system.[5] Ptolemy's *canones* themselves were not explicitly mentioned until the end of the tenth century, when Lupitus of Barcelona referred to them briefly in his introduction to a treatise on the astrolabe. It is only in the eleventh century that we see the first real familiarity with them, when the author of *On the Uses of the Astrolabe* recommended their treatment of the varying length of the day.[6]

If the *Preceptum* stems from the sixth century, we must ask why Ptolemy's astronomy went almost unregarded for over four centuries. An easy answer would be that Ptolemy's astronomy was beyond the capabilities and needs of sixth-century Roman learning, and this would certainly be true for the full theoretical understanding of Ptolemy. Yet the *Preceptum* is a practical handbook, with step-by-step instructions for computing the positions of the Sun and Moon, the time of solar and lunar eclipses, and the degree and sign of the zodiac on the horizon (the *horoscopus*) and the degree at mid-heaven using data from the accompanying tables. The scope of the *Preceptum* is limited; it omits

3. Cassiodorus, *Institutiones*, 2.7.3.
4. Pingree, "Preceptum canonis Ptolomei"; Pingree, "Boethius' Geometry and Astronomy," p. 159; Noel M. Swerdlow, personal correspondence, 14 January 1991.
5. Bede, *De temporum ratione*, 11.70–91, 14.24–6; Hrabanus Maurus, *De computo*, 30.
6. Honigmann, *sieben Klimata*, pp. 102–107; Lupitus of Barcelona(?), *Fragmentum de astrolabio*, ed. Millás Vallicrosa, p. 275.121, ed. Bubnov, p. 375; *De utilitatibus astrolabii*, 1.2, 13.2.

instructions and tables for computing the positions of the five planets.[7] The calculations it does include could help determine the date of Easter or assist stellar timekeeping or astrology, yet we see no signs of its use. Since these calculations required only simple arithmetical manipulation, not complex geometrical demonstration, something other than technical complexity must explain the *Preceptum*'s lack of influence.

One possible factor is the imperfect assimilation of Greek astronomy displayed by the Latin text of the *canones*. There was a limited Latin technical terminology which the translator could have used, yet he ignored these existing Latin terms and filled his translation with Greek terms transliterated into the Latin alphabet. Where Pliny and Martianus Capella had written in ordinary Latin of the obliquity (*obliquitas*) of the circle of the signs (*signifer*), our translator used the Greek *loxoseos*, inclined circle, for the ecliptic without further clarification. Discussing lunar eclipses he produced a sentence that was literally Greek to his readers; if we translate his Latin into English, but leave the Greek, it reads: "A lunar eclipse surely occurs when the *apocrysis* occurs near the *anabibazon* or the *catabibazon*."[8] Given the declining knowledge of Greek in late antiquity, a text of this sort could do little to preserve, much less to disseminate, astronomical knowledge in the Latin West. Such unexplained technical jargon in the only early Latin Ptolemaic text may help explain the minimal influence of Ptolemy's astronomy from the presumed translation of the *Preceptum* in the sixth century to the translation of Arabic astronomical works in the eleventh and twelfth.

If this were not enough of an obstacle, the calendric complications arising from the tables' use of Egyptian reckoning and the conversion from the Alexandrian to the Egyptian reckoning used in Ptolemy's astronomical tables added a further – and not fully explained – complication to what should have been relatively simple calculations.[9] Only those tables that did not rely on the date but depended on the position of the Sun could be easily used. Significantly, these were the only tables commended by the eleventh-century author of *On the Uses of the Astrolabe*. In sum, the *Preceptum* was the sole example of astronomical calculations based on geometrical models to survive in the Latin West, yet it remained a curiosity of literary interest, written by Ptolemy the founder of astronomy, cited as important by Cassiodorus, but scarcely used and scarcely intelligible.

7.  Ptolemy, *Tables manuelles*; Theon of Alexandria, *"petit commentaire,"* pp. 228–235, 239–244, 315–319, 320–322; *Preceptum canones Ptolemaei*, BL MS Harley 2506, ff. 70r –73v. Professor Noel Swerdlow graciously provided me with a xerographic copy of this manuscript.

8.  BL MS Harley 2506, fol. 60v.

9.  Raymond Mercier considers this to have made the Ptolemaic tables useless for astronomical calculations. Raymond Mercier, "Astronomical Tables," p. 115.

## Late Roman learning

Loss of contact with the computational techniques of geometrical astronomy came at the end of a long decline of interest in technical astronomy. Within the Roman tradition philosophy had been taught to the children of the aristocracy, to the young men who would be expected to take up their roles in government. Thus a recurring theme was the Platonic dictum that a society is governed well when philosophers rule and rulers philosophize.[10] Since education aimed to form the leaders of the community rather than expert astronomers, astronomy was presented without technical details. In fact, to the extent that students might have found them unpalatable, too many details would have been a positive hindrance to the main educational goal. Thus there was that almost inevitable simplification which we see today in courses of "science for poets." Yet late antique authors frequently went beyond this to what we might call "science by poets," in which literary presentation was more important than rigorous demonstration, philosophical significance more important than mathematical precision.

Cicero (106–43 B.C.), who had translated Aratus's *Phaenomena* into Latin, considered poetic skill more important than technical knowledge, noting that Aratus had composed a poem on the heavens though he knew no astronomy.[11] Aratus was not unique; the same pattern appears in the works of astronomical poets from Hesiod to Virgil. Cicero himself inspired one of the most peculiar literary presentations of astronomical lore. At the conclusion of his dialogue, *The Republic*, one of its participants, Scipio Africanus the younger, describes a dream in which he ascended into heaven and met his grandfather. The elder Scipio showed his grandson the heavenly spheres, compared to which the earth and the Roman Empire dwindle to insignificance. The uniformity of their motions, the musical harmonies that bind them together, and the insignificance of a human lifetime in comparison to the Great Year in which all the planets return to the same place, all teach that we should value only eternal rewards. Such rewards, the elder Scipio said, are gained through service to the republic. Cicero's goal was not to teach astronomy as much as to use cosmology to illustrate the value of the immortal soul and its eternal celestial dwelling place, in contrast to the transitory terrestrial world of material things.

Such a cosmological presentation openly invited technical commentary, which was provided by Ambrosius Theodosius Macrobius (ca. 360–post 422). His commentary was widely read. More significantly, its astronomical portions circulated independently during the Middle Ages.[12] Macrobius's commentary is true to the

---

10.  Boethius, *De consolatione philosophiae*, 1, pros. 4; Isidore of Seville, "Institutionum Disciplinae."
11.  Cicero, *De oratore*, 1.16, transl. E. W. Sutton and H. Rackham (Cambridge: Harvard Univ. Pr., 1948).
12.  Macrobius, *In Somnium Scipionis*; Stahl, *Macrobius' Commentary*, pp. 39–51; *Roman Science*, pp. 151–169.

Table 4. *Macrobius's planetary distances*

| Planet | Distance from Earth |
|---|---|
| Moon | 1 |
| Sun | $1 \times 2 = 2$ |
| Venus | $2 \times 3 = 6$ |
| Mercury | $6 \times 2^2 = 24$ |
| Mars | $24 \times 3^2 = 216$ |
| Jupiter | $216 \times 2^3 = 1728$ |
| Saturn | $1728 \times 3^3 = 46656$ |

spirit of his source, for both consider the place of the cosmos within Neoplatonic philosophy to be more important than any detailed exposition of astronomical theory. Yet Macrobius did present the general Greek model of the spherical earth at the center of the seven celestial spheres, the outermost of which is girded by ten circles: the ecliptic, the equator, the two tropics, the arctic and antarctic circles, the two colures (passing through the poles and the points of the solstices and the equinoxes), the meridian, and the horizon.[13]

To reinforce Cicero's thesis of the unimportance of earthly things, Macrobius reiterated the astronomers' assertion that the earth is a mere point in comparison to the whole cosmos, providing crude computations of the size of the earth, of the Sun's orbit, and of the Sun itself. But he was more a philosopher than an astronomer. He gave instructions for computing the distances of the planets as in Table 4, interpreting an obscure passage in the *Timaeus* where Plato had related the spheres to the numbers two and three and their squares and cubes.[14] These distances are derived purely from mathematical speculation; they bear no relationship to the distances Ptolemy derived in his *Planetary Hypotheses* by applying an astronomical model to a few imprecise measurements.

In his discussion of the order of the planets Cicero had alluded to their astrological significance, describing Jupiter as propitious and helpful or Mars as dreaded on Earth, which led Macrobius into the realm of astrology. Macrobius maintained that only the Sun and the Moon can have any influence on our lives, for (following Plato) sense perception comes from the Sun while growth comes from the Moon. He then quoted Plotinus's theory that the planets are not causes, but are only signs of an established order, just as birds through their flights and cries unwittingly indicate future events. "And so we have good reason to call this planet

---

13.  Macrobius, *In Somnium Scipionis*, 1.15.8–19.
14.  Macrobius, *In Somnium Scipionis*, 1.20.11–31; 2.3.14–15; cf. Plato, *Timaeus*, 35B –36A. The numbers in the table are not given in Macrobius's text, although he gives instructions from which they could readily be computed.

beneficial and that one baneful since we obtain premonitions of good or evil through them."[15]

Like other Roman encyclopedists, Macrobius was a compiler. He presented plausible statements to support his philosophical view of an orderly cosmos guided by a divinely established mathematical harmony, but many of his statements will not stand up to close technical scrutiny. He maintained that the Egyptians divided the zodiac into twelve equal parts by noting the stars that were rising when two hours had passed, as measured by a water clock.[16] This account cannot apply to the signs of the zodiac; astronomers had already demonstrated geometrically that each sign takes a different time to rise and that these rising times vary at different latitudes. Macrobius does not seem concerned with such astronomical details, nor do his readers. But there is something about a literary tradition in which a treatise on politics required an astronomical commentary, that insured astronomy its continued place in the ideal of a liberal education. When after a long twilight that ideal was revived, astronomy would play a significant part in its revival.

One book that preserved the link between astronomy and philosophy was Chalcidius's demanding commentary on Plato's *Timaeus*, written early in the fourth century. Chalcidius's obscurity stems from his closeness to the tradition of Greek astronomical handbooks, not from any real originality. His astronomical section is a free translation, often retaining bits of Greek technical jargon, from Theon of Smyrna's *Manual of Mathematical Knowledge Useful for an Understanding of Plato*, which in turn is an adaptation of an earlier work by Adrastus.[17] Plato's *Timaeus* was the most "physical" of Plato's books, focusing on the mathematical basis of the universe and the creation of the cosmos by a divine craftsman. This notion of a divine creator was one of the elements in Plato's thought that had appealed to early Christian writers, making Plato their philosopher of choice. The *Timaeus* itself was the sole original Greek philosophical work available in the early Middle Ages, yet it became influential only with the ninth century's renewed interest in Greek learning, stimulated by the philosopher and translator John Scottus Eriugena. Only then did Chalcidius's commentary begin to exercise significant influence.[18]

Chalcidius's commentary covers many standard topics of an astronomical handbook. He discussed the sphericity of the earth and the universe, the circles that gird the cosmos (the equator, the tropics, the arctic and antarctic circles, the zodiac, the horizon, and the meridian). He placed the spheres that carry the planets in order, noting that Eratosthenes and Plato differed from the "Pythago-

15.  Macrobius, *In Somnium Scipionis*, 1.19.19–27.
16.  Macrobius, *In Somniun Scipionis*, 1.21.12–21.
17.  Stahl, *Roman Science*, pp. 143–144; Neugebauer, *HAMA*, pp. 630, 949–950, 958.
18.  Raymond Klibansky, *The Continuity of the Platonic Tradition* (London: The Warburg Institute, 1950), pp. 21–25.

reans" by placing the Sun immediately above the Moon, followed by Mercury, Venus, and the other planets. His explanation of solar and lunar eclipses is conventional, including a discussion of the relative sizes of the earth, the Moon, and the Sun derived from the geometry of their shadows and their presumed distances.

Chalcidius presented the apparent nonuniformity of the motions of the planets more clearly than other Roman handbook writers. He began by noting that their apparent inconstancy is not real, for there is no inconstancy in divine actions. The planets actually move equally, so that they cover equal spaces in equal intervals of time; it only seems to us that they do not move equally. Working with this assumption his discussion proceeds logically from reported observations to geometric models that explain those observations.

He began with the Sun, whose motion is simplest. He noted the ancient observations — observations by Hipparchus and Ptolemy that were clearly beyond his ability — that some seasons are longer than others. Thus the Sun seems to move through the four equal parts of the universe in unequal times, taking 94½ days from vernal equinox to the summer solstice, 92½ to the autumn equinox, 88⅛ days to the winter solstice, and 90⅛ days to return to the vernal equinox. He explained this apparent nonuniformity by two possible mechanisms: an eccentric or an epicycle. He showed, in a qualitative rather than a rigorous sense, how under either model the Sun will seem to move slower in Gemini and faster in Sagittarius. If he did not provide a full geometrical demonstration of how an eccentric or epicycle can account for the unequal seasons, he certainly provided a clear outline for such a demonstration.[19]

Chalcidius extended his commentary to touch on Ptolemy's epicyclic model for the planets, asserting that the varying sizes of their circles and speeds of their revolutions can explain the apparent diversity of their motion and even the periods when they reverse their motion through the stars. His commentary is not always clear, nor even always correct, yet he presented astronomy as a logical mathematical study, the knowledge of which he saw as an essential element to an understanding of Platonic cosmology. This approach to Plato assured astronomy's time-honored intellectual niche: that of a prerequisite to philosophical study.

The most widely influential general introduction to the liberal arts was the fanciful allegory *On the Marriage of Philology and Mercury*, written by the North African writer Martianus Capella (fl. ca. 410–439) for his son.[20] The allegory centers on a wedding where Mercury, the bringer of learning, presents to his bride seven handmaids, representing the seven liberal arts. Each of these, in turn, gives an account of her subject, generally concluded with a poetic epilogue. Like many

19.  Chalcidius, *In Timaeum*, 78–82.
20.  Many of the facts of Martianus's life are uncertain; even his dates may be a century earlier. See Stahl, *Martianus Capella*.

of his contemporaries Martianus drew, often without acknowledgement, from his predecessors. In many cases he did not fully comprehend the point being made; an unaided reader could not master the liberal arts from Martianus alone. Yet he did provide an introduction in an attractive package. Embellished with poetry, and cast in a literary form that mirrored the marriage of eloquence and learning, Martianus's allegory sweetened the learning that Romans were increasingly finding bitter and unpalatable. Its importance as an introduction to the arts, and especially to the art of astronomy, was recognized from Gregory of Tours in the sixth century to scholastic commentators at the beginning of the thirteenth.[21]

The heavens were presented by Astronomy, and to a lesser extent by her sister Geometry, in compact, thorough, and well-organized accounts. Astronomy bears in her hands a measuring instrument and a book containing the predestined paths of the planets and the recurring courses of the stars. Geometry is garbed in a starry peplos and carries a surveyor's instrument and a celestial globe, since she measures both the heavens and the sphere of the earth, which rests in the center, its size like a point when compared to the rest of the universe. Astronomy's realm is governed by measure and number, but there is little mathematical detail in the sisters' accounts. General descriptions of celestial circles, even vaguer allusions to the planets' epicycles and apogees, approximate values of planetary periods, and discussions of the changing length of daylight are as far as Astronomy goes into the world of figure and number. Astronomy tells her medieval audience little about prediction; more could be found by reading the computists. But she has much to offer her listeners in the way of the general concepts of astronomy and a technical vocabulary in which those could be expressed.[22]

When Astronomy ventures from generalities to specifics, she reveals Martianus's limited understanding of his sources. At one point she recounts the time it takes each sign of the zodiac to rise, noting that the varying risings of the signs underlie the changing duration of day and night. Later she tells how the length of daylight varies month by month through the year. The numerical values in these two different accounts are totally inconsistent, and it is quite apparent that Martianus did not know how to reconcile them.[23] Martianus's technical limitations are further shown in Astronomy's account of the Sun's motion across the sky; she erroneously states that the Sun travels the same daily path when it is at opposite points on the zodiac, a claim which is patently untrue as would be made immediately apparent by considering the different paths the Sun takes at mid-summer and mid-winter.[24]

21. *Causa efficiens [astrologiae]: Martianus*. Anonymous, *Accessus philosophorum. vii. artium liberalium*, in Lafleur, *Quatre introductions*, p. 219; Gregory of Tours *Hist. Franc.*, 10.31.

22. Martianus Capella, *De nuptiis*, 8.811.

23. Martianus Capella, *De nuptiis*, 8.844–6, 878; Neugebauer, *HAMA*, pp. 723–724.

24. Martianus Capella, *De nuptiis*, 8.872. This may be a garbled account of the astrological doctrine of reflections in which signs equidistant from the tropics "see" each other. Tester, *A History of Western Astrology*, pp. 140–141.

Astronomy does tell her listeners a few intricate details about her realm; she notes that the Sun takes thirty-two days to pass through Gemini but only twenty-eight days to pass through Sagittarius, explaining this nonuniformity by the fact that the Sun's orbit is eccentric to the earth. But it is by no means clear from Astronomy's account whether the effect is an appearance of slowness due to the greater distance or the physical effect of climbing more slowly to the greater height.[25]

As in many late antique works, there are allusions to the philosophers but not always with full understanding. Astronomy works in a Platonic world of immaterial ideas; she stresses that she speaks of ideal circles and spheres rather than the bronze armillary spheres that mortals use to assist them in comprehending the heavens. Yet she does not mention this as Plato's doctrine, nor when speaking of the peripatetics do any of Martianus's characters give a hint that the peripatetics are followers of Aristotle. If Martianus knew, he certainly did not tell his readers.[26]

## Antique learning in Ostrogothic Italy

As Martianus was finishing his work, the Roman Empire was slowly disintegrating. The Vandals overran his beloved Carthage in 439; the last emperor in the West gave up his nominal power in 476, and before the end of the century, in 493, the Ostrogothic leader Theodoric had established an independent Ostrogothic kingdom in Italy. Roman government was effectively at an end; the attenuated murmurs of Roman learning and culture which continued to echo from generation to generation became fainter and less distinct with each repetition. Yet Theodoric, who had received some education in Constantinople, valued Roman ideals of civility and learning. Cassiodorus described Theodoric's grandson, Athalric, as studying the course of the stars and other natural phenomena – as a virtual philosopher in regal purple.[27] Even if the reality was less than this courtly flattery suggests, the ideal of the learned ruler was employed effectively in Ostrogothic Italy.

The same ideal of the learned ruler appears in the *Consolation of Philosophy*, written by one of the two aristocratic scholars who embarked during Theodoric's reign on a program to insure that ancient Greek learning would continue to speak to the Latins. Ancius Manlius Severus Boethius (ca. 480–524), who served as consul under Theodoric, wrote the *Consolation* while in prison under a charge of conspiring to overthrow Theodoric. The *Consolation* was his greatest achievement, and it continued to provide intellectual stimulation and solace through the Middle

25. Martianus Capella, *De nuptiis*, 25.849.
26. Martianus Capella, *De nuptiis*, 8.815–16, 853.
27. Cassiodorus, *Variae*, 9.24.8.

Ages, but Boethius also wrote on theology, on logic, and on much of the quadrivium.[28]

The *Consolation* is set in prison, where Boethius waited his fate on the charge of conspiracy. Lady Philosophy appeared before him, offering consolation in his ordeal. She reminded him, using the theme we have already seen in Cicero's *Dream of Scipio*, that, just as the astronomers have shown the earth to be an insignificant point, so too are earthly fame and suffering of no real significance. The geometrical model of the cosmos illuminated for Boethius, as it had for Cicero, the intrinsic value of public service. Boethius recalled the words of Plato, "that commonwealths would be good if wise men were appointed rulers, or if those appointed to rule would study wisdom."[29] But Boethius did not depend solely on the philosophers. There are Christian as well as Platonic overtones when Boethius praised the "Creator of the star-filled universe, . . . you rapidly whirl the heavens and give the stars their laws."[30] Boethius's moral lessons impressed the image of the tiny earth, resting like a point at the center of an intelligible and orderly spherical cosmos, upon the minds of medieval thinkers.[31]

Boethius's cosmological image was complemented by the *Introduction to the Divine and Human Readings* of his younger contemporary, Cassiodorus Senator. Sometime after the year 551 Cassiodorus, who had served as consul and master of the offices under Theodoric, retired to a monastery on his estate in southern Italy where he wrote the *Introduction* as a guide for monks. The most significant aspect of Cassiodorus's treatise is that he supplemented his brief summaries of each topic with lists of the books one ought to read to deepen one's knowledge. Since Book Two on secular literature followed the late antique outline of the seven liberal arts, it included a brief chapter on astronomy.

Cassiodorus defined astronomy as the law of the stars, in which their orderly movements and arrangement reflected the decrees of their creator. His discussion presents little in the way of geometrical models, mentioning celestial spheres, directions, and movements in only the most general sense. His most important contribution is to the preservation of Ptolemy's reputation as an authority in astronomy, rather than any profound teaching of his own.

### Astronomy in the Visigothic court

The last compiler of late antiquity, or more properly the first of the Middle Ages, was the Spanish bishop Isidore of Seville (ca. 570–636). Again we see the cultural

---

28. Gibson, *Boethius*. Pingree could not identify any of the surviving texts on astronomy or geometry as indisputably by Boethius. Pingree, "Boethius' Geometry and Astronomy."
29. Boethius, *De consolatione philosophiae*, 1, pros. 4.
30. Boethius, *De consolatione philosophiae*, 1, vers. 5.
31. Boethius, *De consolatione philosophiae*, 2, pros. 7.

interplay in which Roman aristocrats provided philosophic support for the dominion of Germanic rulers. Isidore himself was descended from an old Hispano-
Roman family that, with the decline of Roman political institutions, now exercised its influence through the church. Isidore presented copies of his two principal works dealing with natural phenomena, *On the Nature of Things* and the
*Etymologies*, to the Visigothic king, Sisebut.[32]

Isidore's works contain little of any originality; he was first and foremost an
encyclopedic compiler, yet while he was not presenting a coherent system of the
world, his collections were not just assemblages of isolated "facts" thrown together without theme or purpose. Around the year 613, apparently stimulated by
the lunar and solar eclipses visible in Spain in the previous years, King Sisebut
asked Bishop Isidore to explain the causes of natural phenomena to him. In reply
Isidore wrote the treatise *On the Nature of Things*, in which he urged the king to
avoid superstitious "learning" but rather to follow sound and sober teachings.
Speaking of the special value of knowledge to kings, Isidore added the Old Testament ideal of Solomon to the traditional Platonic ideal of a philosopher-king.
He quoted King Solomon's thanks to God for giving him "true knowledge of
the order of the heavens and the powers of the elements, the divisions of time,
the course of the year and the configuration of the stars."[33] Wisdom would serve
as a means to combat pagan superstition.

King Sisebut responded in kind to Isidore's dedicatory letter with a poetic
explanation of eclipses that, like Isidore's text, stressed that both lunar and solar
eclipses were natural and did not respond to charms or incantations.[34] In this
interchange we see the interaction and mutual support of three concepts of a
single lawful order: the philosophical order of ancient astronomy, the Christian
order of divine law, and the political order of Visigothic political theology, which
sought to portray the Visigothic kings as God's Vicars on Earth.[35]

*On the Nature of Things* is the shorter of Isidore's two scientific works, and its
scope is much narrower than the title suggests. It mainly concerns things in the
sky: astronomical and meteorological phenomena. Isidore spoke of the lengths of
years, the ordinary civil year and the leap year, the ordinary lunar year of twelve
months and the embolismic year of thirteen lunar months. He mentioned the

---

32. Riché, *Education and Culture*, pp. 258–259.
33. Isidore of Seville, *De natura rerum*, praef. 1–2. On the circumstances surrounding the work see
the introduction to Fontaine's edition, pp. 1–6. Isidore quoted only those portions of Solomon's
prayer (Wisd. 7:17–21) that correspond to the contents of *De natura rerum*.
    Plato's ideal, as well as Solomon's, was known in Spain. A text attributed to Isidore tells a
noble Visigothic youth that Plato taught "the commonwealth will fare well when philosophers
rule and rulers philosophize." Isidore of Seville, " 'Institutionum Disciplinae,' " p. 427.
34. Sisebut, *Ad Isidorum*, 18–25.
35. Joscelyn N. Hillgarth, "Historiography in Visigothic Spain," pp. 261–311 in *La storiografia alto-
medievale, Settimane di studio del Centro italiano di studi sull'alto medioevo, XVII* (Spoleto, 1970), at
pp. 280–287, repr. in Hillgarth, *Visigothic Spain, Byzantium, and the Irish* (London: Variorum
Reprints, 1985).

nineteen-year luni-solar cycle, but unlike the computists he was not concerned with the details of computing a calendar. He also listed the names of the months; gave the names of the planets, the order of their spheres, and the periods in which they return to the same place; and explained the planets' retrograde motions and eclipses of the Sun and Moon in very general terms.[36] Isidore's presentation is not always clear; at some points he, or a subsequent scribe, ignored the geometrical framework that could provide his reader a fuller understanding of the point under discussion. When Isidore described the five zones of the earth, the arctic and antarctic, the two temperate zones, and the equatorial region, the accompanying diagram (Fig. 18), or *rota*, places the five circles like five petals on a flower rather than mapping the five parallels on a spherical earth.[37]

Astronomy played a proportionately lesser role in Isidore's encyclopedia, the *Etymologies*. Here Isidore followed the practice of such encyclopedists as Pliny, dispersing astronomy throughout the book under several topical headings. This division of what we would gather under the single heading of astronomy, reflects the way Isidore and his contemporaries organized their understanding of the world. The greater part, dealing with the celestial spheres, the Sun, Moon, and other planets, the constellations, and their motions, appears among the four arts of the quadrivium. Yet Isidore's discussion of time – of days, months, and years and of the solstices and equinoxes – is in a section on chronology that immediately follows a section on law, while a brief discussion of the astronomy underlying the date of Easter is included in a section on books and on the offices of the church. Going beyond astronomy proper, Isidore warned against the superstitions of astrologers in a section devoted to a wide range of religious and philosophical sects.[38]

At no point, however, did Isidore go into great detail; he presented only the merest sketch of ancient learning, which needed to be fleshed out from other sources. Even more significantly, he did not restrict himself to the framework of Martianus Capella and Cassiodorus, who had placed their discussions of astronomy among the mathematical disciplines of the liberal arts.[39] Isidore did not see astronomy just as a single intellectual discipline; it was for him also a series of solutions to a range of practical questions. The astronomies of the early Middle Ages were typified, in structure and in content, more by the astronomical parts of Isodore's *Etymologies* than by the late antique writings on the liberal art of astronomy.

36. Isidore of Seville, *De natura rerum*, 6.1–4; 4.1–6; 23.1–4; 22.3; 20–21.

37. Isidore of Seville, *De natura rerum*, 10.1–3, ed. Fontaine, p. 210bis (fig. 3). There is no reason to think this *rota* was intended as a map, since it would be inconsistent with Isidore's view that the earth is a globe. Stevens, "Figure of the Earth."

38. Isidore of Seville, *Etymologiae*, 3.34–61, 5.27–36, 6.17, 8.9.22–26.

39. Isidore was certainly aware of this classification, which he mentioned in the *Etymologiae* and, with substantial changes, in his *Differentiae*. Díaz y Díaz, "Arts Libéraux Espagnols et Insulaires," pp. 38–39.

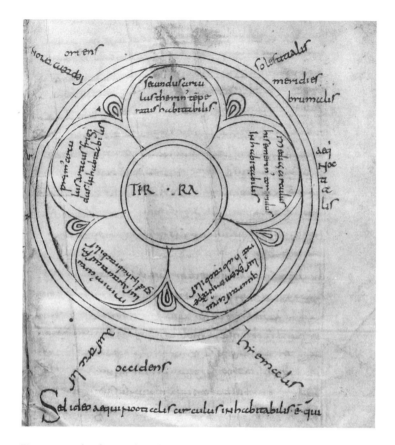

Figure 18. The five circles of the world. Isidore of Seville, *De natura rerum*. Bern, Burgerbibliothek, MS 610, fol. 20r, ninth century. Used by permission.

*Reading clockwise from the ten o'clock position:*

*First circle,* arctic, frigid and uninhabitable.

*Second circle,* earthly, temperate and habitable.

*Third circle,* equatorial, torrid and uninhabitable.

*Fourth circle,* beyond the equator, temperate and habitable.

*Fifth circle,* antarctic, frigid and uninhabitable.

A medieval reader studying those works of late antiquity, especially one seeking in them an introduction to astronomy, would find them to be strikingly incomplete. But they were not written as self-sufficient introductions to technical astronomy. The geometrical structure and size of the cosmos, its diverse parts, and the principles governing their motions were presented to form human character by making the readers aware of their place in the cosmos and conscious of the fact that the cosmos was governed by law. When accompanied by other works, or when taught by a widely read instructor, the late antique works could provide easy, and even pleasant, introductions to further study in astronomy. But in the early Middle Ages these were not introductions; more advanced texts providing computational techniques based on this geometrical structure were lost, and only the structure and its figurative meanings remained as the sum total of classical astronomy. That changed context, as much as any limitations of these texts, accounts for the long neglect of the classical tradition of astronomy.

# PART THREE

# The harvest of medieval astronomies

# CHAPTER EIGHT

# *The fusion of astronomical traditions*

> Let schools be established where children can learn to read. Carefully correct the psalms, notation, chant, computus, grammar, and the Catholic books in every monastery and diocese, because often some desire to pray to God properly, but they pray badly because of the uncorrected books.
>
> Charlemagne, *Admonitio generalis* (789)[1]

The ascent of the Carolingians to the Frankish monarchy – and ultimately to imperial rank – marked a brief phase of consolidation and stability in Europe. One major result of that stability was the intellectual renewal we call the Carolingian Renaissance. In Charlemagne's *Admonitio generalis* we see both the motives and the results of that reform, themes often repeated and elaborated in other decrees of king, emperor, and church councils. From the outset the primary impetus for reform was religious; Charlemagne insisted here that rituals be conducted properly. This, in turn, reflects the governing characteristic of Carolingian reform: uniform adherence to authoritative standards. These concerns called forth two related developments that influenced the development of astronomy. First, schools were established, both elementary schools and, more to the point, higher schools in which the clergy were taught computus among a range of required subjects. Second, and perhaps less obvious, complete and careful exemplars of authentic texts were collected, copied, and disseminated.

Guiding and staffing Charlemagne's educational program was a community of scholars, drawn to the court at Aachen from all the parts of Europe. The most significant of these scholars for the study of astronomy was Alcuin of York (ca. 730–804). Alcuin had been master of the cathedral school of York, where that distinctly Irish and English scholarly program with its emphasis upon the knowledge useful for the reckoning of times (i.e., computus, cosmography, and astronomy) flourished.[2] Recognizing his intellectual debt to Bede, Alcuin included this Northumbrian monk in his poem in praise of the bishops, kings, and saints of York. Although Alcuin left only one astronomical work, a minor work on com-

---

1. *Admonitio generalis*, cap. 72; *MGH*, Leges., Cap., I, p. 60. Riché sees a broader Carolingian concern for education in this text. Riché, *Écoles et enseignement*, p. 71, n. 10.
2. Riché, *Education and Culture*, pp. 384–393.

putus, the frequent allusions to astronomy in his correspondence and other writings reveal his interest in the subject.[3]

Like other medieval clerics, Alcuin was not tied down to one place but traveled much more than we might expect; in 780, on one of his journeys to Italy, he met Charlemagne. In 793 Charlemagne called Alcuin to join the multinational circle of scholars at the "palace school," where he remained until 796, when Charlemagne appointed him abbot of St. Martin's monastery at Tours. The "palace school" was not a school in the usual sense of the word. The circle of scholars surrounding Charlemagne taught the children at court, wrote poetry, advised Charlemagne on matters of state and church, discussed scholarly questions with him and other members of his court, and obtained books and had them copied for Charlemagne's library.[4]

To insure the accurate copying of texts a new, more compact, and more legible form of handwriting known as Caroline minuscule was established as a uniform standard for the scriptoria of the empire. To insure that the copies were authentic, standard versions were made from the oldest and best texts, from which reliable copies could be made for monasteries, churches, and schools. Authentic texts were provided for the whole spectrum of Carolingian learning. In 781 Pope Hadrian sent Charlemagne the Gregorian sacramentary, a book of church rituals as performed at Rome; Charlemagne had copies made to insure conformity in prayer throughout his kingdom.[5] In 787 Charlemagne had seen an early copy of St. Benedict's monastic rule at Monte Cassino; he ordered that a copy be made for his palace at Aachen so monasteries would follow an authentic rule.[6] Alcuin of York presented Charlemagne with an ornate copy of his new standard text of the Latin Bible to celebrate the emperor's coronation on Christmas Day, 800.[7] Charlemagne gathered secular as well as sacred works, obtaining copies of literary classics and practical manuals on subjects ranging from agriculture to architecture.[8] For our purposes the most significant Carolingian texts are two related astronomical and computistical anthologies compiled around the year 809.

Many of the astronomical traditions we have seen converged on the vibrant community around the Carolingian court. Alcuin's ongoing correspondence with Charlemagne over astronomical problems sketches the outlines of Carolingian astronomy. Problems of the calendar and of the motion of the Sun and Moon through the zodiac recur frequently. Alcuin's concerns are not strictly astronomical; issues of religious symbolism and ritual uniformity also arise. The year cannot

3.  Max Manitius, *Geschichte der lateinischen Literatur des Mittelalters*, 1. Teil (Munich: C. H. Beck'sche Verlagsbuchhandlung, 1911), pp. 285–287.
4.  McKitterick, *Frankish Kingdoms*, pp. 160–166; Riché, *Daily Life*, pp. 203–207.
5.  Riché, *Daily Life*, pp. 230–231.
6.  Timothy Fry, ed., *RB 1980: The Rule of St. Benedict* (Collegeville, Minn.: The Liturgical Press, 1980), pp. 105–106.
7.  Smalley, *Bible in the Middle Ages*, p. 37.
8.  McKitterick, *Frankish Kingdoms*, pp. 145–152.

begin in September, following Egyptian reckoning, for darkness is still growing then; rather, it begins at the birth of Christ as the light grows from mid-winter. No matter what the Moon in the sky looks like, the Easter computation must follow the decrees of the Nicean Council to insure that the Resurrection be harmoniously celebrated everywhere on the globe. Alcuin's sources do not go beyond the computists and Pliny, whom he urges Charlemagne to study. Although he mentions the seven liberal arts, Alcuin's version of astronomy strays from that tradition; he considered astronomy as a Pythagorean discipline founded on numerical harmonies, reflecting the computists' emphasis upon arithmetic rather than the antique emphasis upon geometry.

Carolingian astronomy was not mere book learning. Charlemagne's queries reflect his observation of the heavens, and he was concerned when the skies did not conform to calculations.[9] The emperor's biographer, Einhard, tells us that Charlemagne had learned "the art of computus and with great keenness and curiosity investigated the course of the stars."[10] When Alcuin describes how one of Charlemagne's daughters "would study the stars of heaven at night, and was accustomed to ceaselessly praise the powerful God who adorns the firmament with constellations, the earth with greenery, who by His Word makes all the wonders of the world," we see that the psalmist's emphasis upon the worship of God through his Creation flourished at court.[11]

Charlemagne continued his interest in astronomy after Alcuin's death, asking the Irish monk Dungal of St. Denis about two eclipses of the Sun reported to have occurred in the year 810. Dungal's reply gave a summary of the circles and spheres of the heavens and the motions of the Sun, Moon, and other planets taken from Macrobius's *Commentary on the Dream of Scipio*; a brief discussion of the interval between successive eclipses taken from Pliny's *Natural History*; and a discussion of the Great Year of fifteen thousand years, again taken from Macrobius. Dungal used the Great Year to claim for astronomers the ability to predict eclipses and other celestial events not only a month in advance, or a year, or twenty years, or a hundred or a thousand, but over fifteen thousand years.[12]

Dungal himself, knew of only one of the eclipses that concerned Charlemagne, the nearly total eclipse on 30 November. But given the date of the observed November eclipse, Charlemagne's query about an earlier eclipse in that same year, and Pliny's account of the interval between eclipses, he calculated that an eclipse must have occurred six months earlier on 7 June. Pliny's interval showed that

9.  Alcuin of York, *Epistolae*, 126, 145, 148, 149, 155, 170, 171; Dietrich Lohrmann, "Alcuins Korrespondenz mit Karl dem Grossen über Kalender und Astronomie," pp. 79–111 in Butzer and Lohrmann, *Science in Carolingian Times*.
10.  Einhard, *Vita Karoli Magni*, 25.
11.  Alcuin, *Carminae*, 26.41–44.
12.  Dungal, *Epistolae* I, MGH, Epp., 4, pp. 570–578; cf. Pliny, *Hist. nat.*, 2.10.57. Eastwood, "Dungal's Letter to Charles the Great."

there was nothing unusual about this repetition of solar eclipses. In fact, the June eclipse did not occur on the date Dungal had calculated and, in any event, could not have been seen from Europe. The only possible second eclipse for that year was a partial eclipse barely visible in high northern latitudes five months earlier on 5 July, an anomalously short interval between eclipses.[13]

This discussion confirms the picture we have from Alcuin of the kind of astronomy practiced in Carolingian court circles. Although Dungal presented the general structure of the heavens, noting that eclipses occur only when Sun and Moon are both on the ecliptic and that at other times the Moon's shadow passes above or below the earth, he did not use the inclined path of the Moon to explain geometrically why eclipses are normally separated by six months. Instead he took the average interval between eclipses from Pliny and used conventional computus to compute the date of the sixth preceding new Moon, and hence of the preceding eclipse. Apparently Dungal could not connect computistical predictions of recurring events based on average intervals in time with the geometrical astronomy of motions in space.

Dungal and his contemporaries did not know the kind of geometry that could have been applied to these problems. Excerpts from Euclid's *Elements*, consisting of definitions, axioms, and postulates from the first four books and selected propositions, only three of which were presented with proofs, had been known since late antiquity.[14] But the works of Theodosius and Menelaus in spherical geometry, which could provide a firm theoretical basis for solving problems in geometrical astronomy, would not be translated until the twelfth century.[15]

Computistical issues were frequently raised at the turn of the century; astronomy was the dominant science of the 800s. In 809 a group of computists was called together to be examined on certain computistical details. The questions reflect the wide-ranging Carolingian concern with ritual uniformity in accordance with authoritative methods and texts, more than they do a dispute over the proper methods of computus such as had arisen in Ireland and England.[16] This examination and the contemporary computistical anthologies seem intended to stop the kind of confusion reflected in a computistical manuscript composed at the end of the eighth century, now in the Cathedral Library of Cologne, which included an

---

13.  Besides the eclipse on 5 July there was also an eclipse on 5 June, but this was not visible in Europe or anywhere in the northern hemisphere. The two eclipses discussed in Dungal's letter are also mentioned in the later seven-book computus and in many Carolingian annals. D. J. Schove, *Chronology of Eclipses and Comets: AD 1–1000* (Woodbridge, Suffolk: Boydell Press, 1984), pp. 176–179; V. King, "Some Astronomical Excerpts," p. 75.

14.  Folkerts, *"Boethius" Geometrie II*; Pingree, "Boethius' Geometry and Astronomy"; cf. Stevens, "*Compotistica et astronomica*," pp. 39–45.

15.  Kibre, "*Quadrivium* in the Thirteenth Century," p. 185, n. 68.

16.  Charles W. Jones, "An Early Medieval Licensing Examination," *History of Education Quarterly*, 2(1963):19–29, gives the text and translation; see also *MGH*, Epp. 4, pp. 565–567.

"extremely corrupt" version of Bede's *De temporum ratione*, a number of early Irish texts on computus, and a lunar table based on the abandoned procedures of Victorius of Aquitaine.[17]

The Carolingian computistical anthologies further recast the study of astronomy by adding new antique texts, reflecting the Carolingian interest in classical science and mythology, to the traditional computistical material. Two different versions of the anthologies are extant: one, of three books, compiled around 810 and the other, of seven books, at Salzburg around 818. They share the same general organization and much of their content, including computations made at Charlemagne's court for the year 809.[18] Each anthology begins with a calendar that lists the feasts of the church; the dates of astronomical phenomena such as solstices, equinoxes, and the entry of the Sun into each sign of the zodiac; and other seasonal phenomena such as weather and the coming of birds.[19] The next section includes computistical tables and texts giving instructions for various calendric computations. The final section shows the influence of the emerging interest in the traditions of ancient astronomy, with a series of excerpts on the structure of the heavens from Pliny's *Natural History* immediately preceded by descriptions of the constellations taken from various ancient sources.

These "star catalogs," as they are commonly called, are a far cry from what modern astronomers and ancients such as Ptolemy meant by that term; they do not give the position of the stars in any mathematical system of coordinates. Rather they are qualitative descriptions of the constellations, noting the number of stars in each part of the constellation and the general location of the brighter stars. These descriptions are accompanied by illustrations, yet the illustrations depict the mythological figure represented by the constellation rather than the location and brightness of individual stars (Fig. 19). In some copies the figures contain no stars at all, the body of the figure being filled instead with the text discussing the constellation. As guides to astronomical observation these "star catalogs" could, at most, indicate the general location of constellations or serve as reminders of the relative positions of already familiar constellations. They are more suited for mythological edification than for astronomical observation.

The final part of these anthologies draws together various discussions of the

17. Wesley M. Stevens, ed., Rabanus Maurus, *De Computo, CCCM* 44, introduction, pp. 171, 175; Borst, *Plinius und seine Leser*, pp. 146–156; Anton von Euw, "Die künstlerische Gestaltung der astronomischen und komputistischen Handschriften des Westens," pp. 251–269 in Butzer and Lohrmann, *Science in Carolingian Times*, here pp. 252–255; Jones, *BOT*, pp. 82, 96, 108, 111, 151, 354.

18. King considers them as descended from a common source, dated on the basis of these shared calculations to the year 809. V. King, "Some Astronomical Excerpts," pp. 57–58, 77–78. For further details of the development of these anthologies see Borst, *Plinius und seine Leser*, pp. 156–164, 171–173; "Enzyklopädie von 809."

19. The contents of the two versions are detailed in V. King, "Some Astronomical Excerpts," pp. 3–22, 28–45, 55–62. The calendar from the "three book computus" is in McCulloh, "*Martyrologium Excarpsatum.*"

Figure 19. The constellation Cygnus. From the Carolingian computistical collection of 809. The artist made no attempt to depict the precise order of the stars. Compare the cruder but more accurate representation of the same constellation in *De cursu stellarum*, where it is called the Greater Cross (Fig. 15). Vatican, Cod. Reg. Lat. 123, fol. 196v. Foto Biblioteca Vaticana. Used by permission.

geometrical structure of the universe found in the liberal arts tradition. Excerpts from Pliny's *Natural History* and Bede's *On the Nature of Things* are common elements of these collections. These texts provide a general cosmological picture, discussing the dimensions of the Sun, Moon, and earth; the spaces between the celestial spheres; the motion of the planets through the zodiac; and the causes of eclipses. As should be expected of excerpts from an encyclopedist like Pliny, the texts do not present advanced astronomical data. The section on the distances of the planets from the earth arranges the planetary spheres according to Pythagorean musical cosmology (Fig. 20), rather than according to any astronomical measurements or theory. The texts are accompanied by figures showing the orbits of the planets around the earth with their spacings, the changing distance of the planets from the earth, and the motion of the planets above and below the ecliptic as

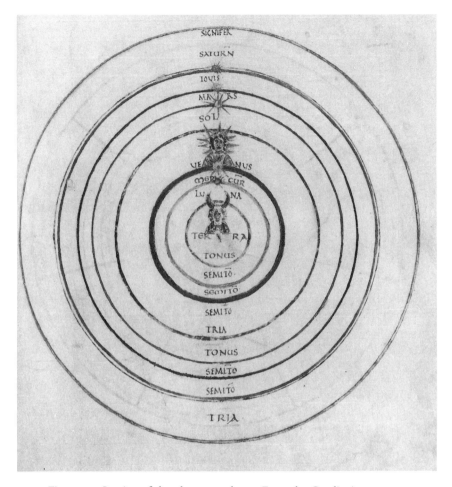

Figure 20. Spacing of the planetary spheres. From the Carolingian computistical anthology of 809. The musical spacings between the spheres are in the
excerpts from Pliny, *Hist. nat.*, 2.20.84.

From Earth to the Moon, a tone; to Mercury, a half tone; to Venus, a half
tone; to the Sun, three half tones; to Mars, a tone; to Jupiter, a half tone; to
Saturn, a half tone; and to the zodiac, three half tones. Vienna, Österreichischen Nationalbibliothek, MS 387, fol. 123r. Used by permission.

they make their way around the zodiac. The figures attempt to represent these
elements accurately, but we would be mistaken to confuse them with graphs of
scientific data. While the figures show the general range of each planet above and
below the ecliptic and the place of each planet's greatest distance from the earth,

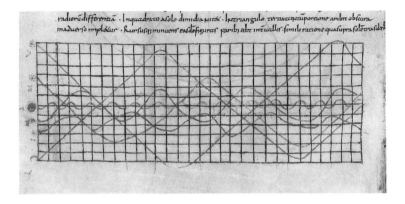

Figure 21. The latitudes of the planets. Diagram of the motion of the planets above and below the ecliptic. The figure accurately depicts Pliny's values for the range of the planets' motion, but the periodicity of the motion is purely arbitrary. Bern, Burgerbibliothek, MS 347, fol. 24v, ninth century. Used by permission.

the apogee,[20] as given in the accompanying texts from Pliny, these are not mathematically precise graphs; they are illustrations to a text. For example, one of the common illustrations that seems to graph the planets' deviation above and below the ecliptic (Fig. 21) completely ignores the periodicity of this motion, the general shape of the latitude curves being determined by considerations of artistic symmetry rather than quantitative accuracy.[21]

But it was not just the geometrical tradition which contributed diagrams to the Carolingian anthologies. The three-book computus contains separate figures depicting the courses of the Sun and Moon through the zodiac. The figures make no attempt to depict the paths of the Sun and Moon; rather, they depict a personification of the Sun or Moon at the center. Around the periphery of the solar diagram (see Fig. 12) are the signs of the zodiac, the months, notations of the solstices and equinoxes, and markings of the mid-quarter days. Noted on radial lines are the duration of daylight and nighttime, computed using a simple linear increase or decrease of daylight by one hour every fifteen days.[22]

An earlier Carolingian computistical manuscript provided a geometrical depiction of the paths traced by the Sun from the places of sunrise to sunset at the

20. Pliny gives two apogees: one is the purely astrological concept of the planet's exaltation, the place where its influence is greatest; the other corresponds to the eccentric apogee of a Ptolemaic model. The diagrams reflect the latter eccentric apogee. The changing distance of the planets from the earth is influenced much more by the epicycle of Ptolemy's model than by the eccentricity. However, only the eccentric apogee is fixed in relation to the zodiac.

21. Eastwood, "Plinian Astronomical Diagrams."

22. Vienna, Österreichische Nationalbibliothek, Cod. 387, ff. 137r, 138r; cf. Munich, Bayerische Staatsbibliothek, CLM 210.

solstices and equinoxes. It is not clear whether the diagram should be read as a projection of the Sun's paths onto a horizontal plane or as a projection onto an inclined equatorial plane. I suspect that the artist, if asked, might have had trouble making the distinction. Nonetheless, the diagram does clearly mark where the Sun rises and sets at the solstices and equinoxes, where it passes overhead at noon, and where it passes below the earth at midnight.[23]

These figures began to transform the way medieval students of computus saw the heavens. Computus had always been an arithmetical art, mainly concerned with computing discrete phenomena: the date of the Paschal Full Moon or the age of the Moon at the beginning of the year or on any other day. Now the astronomy of the liberal arts, with its explicit visualization of the heavens as geometric bodies, as spheres undergoing continual motion, had made its way into the standard corpus of astronomical texts. Computing the slowly changing positions of the stars, Sun, Moon, and other planets could now be seen as essential astronomical problems, problems for which neither the arithmetical techniques of computus nor the surviving remnants of geometrical astronomy provided adequate answers.

The seven-book computus added further selections and paraphrases from Macrobius's *Commentary on the Dream of Scipio* and Martianus Capella's *On the Marriage of Philology and Mercury* that relate the geometric understanding seen in the texts of Pliny and their accompanying illustrations to a renewed recognition of the role of quantitative measurement in astronomy. The first of this series is Macrobius's discussion of how to measure the diameter of the Sun using a spherical sundial, which he finds to be 1/216th of the circumference of the solar orbit, or 140,000 stadia, or twice the diameter of the earth. The next is a related paraphrase from Martianus Capella, telling how to determine the apparent diameter of the Moon by using a water clock to measure the time it takes the Moon's disk to rise. Martianus concludes that the Moon's diameter is 1/600th of the circumference of the lunar orbit. The third element is from Martianus's cryptic discussion of Eratosthenes's measurements of the size of the earth, which he presented as one of the major achievements of geometry.[24]

Such discussions of the dimensions of the earth, the heavenly bodies, and the spheres that carry them around are nothing new. But here we see collected in one place a set of short excerpts showing how measurements, or more properly, purported measurements, may be used to compute these dimensions. There is a recognition of how measurements of time can be related to arcs struck off on a

---

23.  Cologne, Dombibliothek, cod. 83[II], fol. 81v. von Euw, "Die künsterlische Gestaltung," p. 256, Abb. 10; cf. Obrist, "Wind Diagrams," p. 59, figs. 16–17.

24.  V. King, "Some Astronomical Excerpts," pp. 43–44, 72–73. These texts are commonly titled "de mensura et magnitudine terrae et circuli per quem solis iter est" from Macrobius, *In Somnium Scipionis*, 1.20.14–24, 25–32; "De mensura lunae," from Martianus Capella, *De nuptiis*, 8.860; and "Quo magnitudine terrae depraehensa est" from Martianus Capella, *De nuptiis*, 6.596–8.

celestial body's orbit and from there to the actual physical dimensions of the body itself. The notion that time is related to angle and angle to dimension had not played a significant part in computus before. In the Carolingian era astronomy was becoming increasingly geometrical, but this was a geometry in the Capellan sense of measurement rather than the Euclidian sense of geometrical proof.

Finally, along with these new texts of computus, other forms of astronomical texts that seem more decorative than practical began to circulate in Carolingian court circles. They lacked the computistical texts and tables, but they continued to carry the cosmological selections from Pliny, their associated illustrations, and various forms of the "star catalogs." But unlike those of the computus collections, these "star catalogs" were often supplemented by accounts of the constellations' mythological background.[25] The planetary gods were now dead and buried and could be safely exhumed for artistic display and antiquarian study at court.

## Astronomy and court culture

As the Carolingians increasingly looked back to classical antiquity for imperial models and mythic symbols, the ancient claims of astronomy to provide regal symbolism and of astrology to provide celestial foresight began to provide additional motives for the interest in astronomy at the imperial court. One of the earliest and most opulent signs of this courtly astronomy was Charlemagne's famous silver table, made in three parts, on which was a description of the entire celestial sphere, of the stars, and of the courses of the planets. At Charlemagne's death the table was the only item that his son, Louis the Pious, reserved for himself out of the treasure sent to the pope to assist priests, the poor, foreigners, widows, and orphans. We are not told whether it attracted him because of its craftsmanship or its astronomical content, although Louis's known interest in astronomy suggests the latter. After Louis's death in 842, the table was finally broken up and lost to history.

Although the table has not survived we can presume it was not unlike artistic representations of the heavens found in contemporary and late antique manuscripts and inscriptions, all of which tended towards artistic and symbolic, rather than mathematically precise, depictions of the constellations and planets. Yet these figures did place the stars within a geometrical framework of the celestial circles: the parallels from the arctic to the antarctic, the inclined band of the zodiac, or the ecliptic circle touching the two tropics where the Sun reversed its motions. In art, as well as in texts, the Carolingians were coming to visualize the motions

---

25. McGurk, "Carolingian Astronomical Manuscripts," pp. 318–321.

of the heavens in a geometrical model of the celestial sphere and its circles, rather than through the computists' arithmetical prediction of the day of future events.[26]

A superb survival of the blending of art and astronomy is an astronomical manuscript associated with the court of Louis the Pious. The deluxe characteristics of this manuscript, its lavish illumination, the quality of the parchment, all indicate that it was made for someone in the court of Charlemagne's successor.[27] It includes among its many exquisite illuminations a depiction of the configuration of the planets at the full Moon after the vernal equinox in the year 816.[28] The planets were all far enough from the Sun that they could have been observed on that date, so there is no need to posit any means to calculate their position beyond the rudimentary calculations of the position of the Sun and Moon found in contemporary computus texts.[29]

The diagram is not purely decorative; parameters derived from Pliny's *Natural History* are written on the paths of the planets. The diagram gives the period of each planet's orbit around the earth, the position of its apogee and perigee, and the position of its astrological exaltation (following Pliny, this is called the *apsis*, with no explicit astrological connotation). For the superior planets it lists, in relation to the Sun, the place where they become stationary, while for Venus it gives that planet's maximum distance from the Sun.[30]

Perhaps the most meaningful example of the courtly use of classical astronomical symbolism appears in the adoption of the antique custom of stellar ceremonial mantles. Such mantles present the ruler as a Solomonic or Platonic philosopher-king, garbed in wisdom, and assimilate his reign to the reign of God over the universe, which He rules by weight and number and measure.

The earliest surviving mantle (Fig. 22) dates from the reign of Emperor Henry II (1002–24). Henry's mantle is of deep violet silk, embroidered in gold thread with what it labels – like Charlemagne's silver table – as a description of the entire celestial sphere (*descriptio tocius orbis*). The mantle displays an eclectic

26. For a possible reconstruction, see Estrey, "Charlemagne's Celestial Table."

27. Leiden, Universiteitsbibliotheek, MS Voss. Lat. Q. 79. Bischoff et al., *Aratea*; Stückelberger, "Sterngloben und Sternkarten"; C. L. Verkerk, "*Aratea*: A Review of the Literature Concerning MS. Vossianus lat. q. 79 in Leiden University Library," *Journal of Medieval History*, 6(1980):245–287.

28. Mostert and Mostert, "the Leiden Aratea." This date seems preferable to Eastwood's earlier calculation for the configuration of 28 March 579, which implied descent of the astronomical data from an ancient source. I am grateful to Dr. Eastwood for bringing Mostert and Mostert's article to my attention.

29. Cf. Hrabanus Maurus's observation of the planets for 9 July 820; *De computo*, 48.14–19. Eastwood and Pingree postulate calculated positions using the *Preceptum canonis Ptolomei*. Eastwood, "Leiden Planetary Configuration," pp. 2–4; Pingree, "Preceptum canonis Ptolomei," pp. 372–373.

30. Eastwood, "Leiden Planetary Configuration."

Figure 22. The stellar mantle of Emperor Henry II (detail). Symbols of the Sun and Moon are embroidered to either side of the central figure of Christ. Near the bottom is Mary, referred to as *stella maris*, the Star of the Sea. Celestial spheres with the constellations and the zodiac are at the lower corners. Archdiocesan Museum Bamberg. Used by permission.

mixture of religious and astronomical imagery. Near the center are conventional antique personifications of the Sun and Moon, slightly above and on either side of a mandorla containing the figure of Christ in majesty. To the sides of the mandorla are the letters alpha and omega, signifying that Christ is the beginning and end of things. Slightly below are figures of Saint John and the Virgin Mary, here called by her title *stella maris*, Star of the Sea.[31] Further from the center are drawings of the eastern and western hemispheres of the celestial sphere, including the equator, the tropics and the arctic circles, the polar axis, and the inclined circle of the zodiac. The bulk of the mantle is adorned with

31. O'Connor, "Star Mantle"; P. E. Schramm and F. Mütherich, *Denkmale der deutschen Könige und Kaiser*, 1, *Ein Beitrag zur Herrschergeschichte vom Karl der Großen bis Friedrich II, 768–1250*, 2nd ed. (Munich: Prestel Verlag, 1981), pp. 163, 348, pl. 130; L. Grodecki, F. Mütherich, T. Taralon, and F. Wormald, *Le siècle de l'An mil* (Paris, 1975), pp. 269–270; Riché, *Écoles et enseignement*, p. 269. An earlier mantle belonging to Otto III was given to the cloister of St. Alexius in Rome. I examined the embroidery of Henry's mantle through color photographs graciously provided by the Diözesanmuseum Bamberg.

repeated symbols of the four evangelists, personifications of the virtues, and thirty-three depictions of significant constellations. The constellation figures and their accompanying descriptions are taken from the Aratus texts circulating in contemporary "star catalogs." The descriptions mix the astronomical with the mythological: the body of the serpent (*serpens*, the modern Draco) is described as twining between the two northern bears, while Hercules is described slaying the serpent.[32] On the fringe of the mantle is a dedication to Henry, "the ornament of Europe," that calls upon "the King who rules forever [to] increase your reign."[33]

Such stellar mantles have a long tradition. They were worn by ancient rulers and are a common motif in late antique art, often portrayed billowing over celestial deities as varied as the Persian Mithras and the Celtic Epona. Mantles were familiar poetic metaphors for the skies, representing the dark mantle of the night, the mantle with which dawn cloaks the stars, and the celestial vault itself.[34] Martianus Capella dressed several of his celestial characters in starry garb. Geometry, whose rules govern the motions of the heavens, wore a deep purple-and-blue peplos decorated with the celestial circles, the golden orbs of the Sun and Moon among the stars, and figures of many kinds. Most significantly, Jupiter's formal attire, which he wore when he addressed the senate of the gods, included a robe decorated with stars; in his right hand he bore two regal globes, while his left hand rested on a nine-stringed lyre, alluding to the celestial harmonies he plays as universal ruler.[35] Equally familiar would be the description in the book of Wisdom (18:24) of the high priest's, Aaron's, long robe on which was the entire orb of the earth (*totus erat orbis terrarum*). An account of imperial costume (ca. 1030) describes "the emperor's golden mantle, having a zodiac made of gold and pearls and precious stones, in whose fringe are 365 golden bells shaped like orange blossoms and just as many oranges." This fringe of bells matches the description of Aaron's priestly robe in Exodus 28:33, but the cosmic symbolism of the zodiac and the number 365 stems from the astronomical tradition of the liberal arts.[36]

---

32.  O'Connor found illustrations bearing close resemblances to those of the stellar mantle in two ninth-century "star catalogs," St. Gall, Stiftsbibliothek, MS 902, and its copy, St. Gall, MS 250. O'Connor, "Star Mantle," pp. 63–66, 108.

    For examples of the textual sources, see *Baudri Abbas ad Adelam Comitissam*, 599–602 in *Commentariorum in Aratum reliquae*, and *Aratea*, 48–50; *Scholia Basiliensia*, 60.6.14; *Scholia Strozziana et Sangermanensia*, 107.14–17, 116.1–4, 189.6–8 in Germanicus Caesar, *Aratea*.

33.  O'Connor, "Star Mantle," pp. 37–38.

34.  Discussing Boethius's allusion to the sparkling stars that decorate the night, Remigius of Auxerre gives the unexpected gloss that "night is depicted as a starry peplum." Remigius, *In Boethius*, 4 metr. 1, ed. Silk, p. 340.; cf. Sisebut, *Ad Isidorum*, 56–57; Martianus Capella, *De nuptiis*, 2.116, 6.584.

35.  Martianus Capella, *De nuptiis*, 6.580–1, 1.66.

36.  O'Connor, "Star Mantle," pp. 151–152, citing the *Libellus de ceremoniis aule imperatoris* (ca. 1030);

The appearance of the emperor in priestly robes, covered with astronomical symbolism connecting his dominion on earth to the divine reign in heaven, displays the imperial ideology of the Holy Roman Emperors. In the continuing tension between imperial and papal authority Henry had styled himself "servant of the servants of Christ and august emperor of the Romans, according to the will of God and of our savior and liberator." By appropriating the traditional papal title of "servant of the servants of God" and claiming that his dominion came directly from God, Henry was claiming the autonomy of his power over any terrestrial rival. While Pope Benedict VIII seemed to restrict the emperor's power and subordinate it to his own when, during Henry's coronation, he gave him a golden sphere surmounted by a cross as an image of the mundane sphere over which the emperor ruled, court artists portrayed the emperor as crowned directly by God, with angels giving him a lance and a sword. The imperial mantle reflected the same imperial ideology, suggesting by its celestial motif that the emperor, like the Byzantine emperors, acted in the world as the vicar of Christ, the true ruler of the cosmos.[37]

There is one more element in Henry's robe; it has two inscriptions that convey astrological messages, which cannot be traced to the Aratus tradition: "When Scorpio rises, deaths increase" and "This star, Cancer, brings evil things of the world." Next to Cancer is a further inscription, "Here let the astrologer be careful."[38]

At the end of the twelfth century the French poet Chrétien de Troyes would use a similar robe to compare terrestrial and celestial kingship; the king governs his realm by law and reason, just as the universe is governed by the divine reason. In Chrétien's Arthurian romance *Erec et Enide*, King Arthur presented Erec with a coronation robe decorated with the four mathematical arts of the quadrivium. But true to the genre, the magical side of astronomy also appears here; Erec's mantle had been embroidered by four fairies, "with great wisdom and skill." The fairy who embroidered astronomy takes counsel from the Moon and the Sun about what she must do, for they make known to her "whatever was and will be . . . with certainty, without lying and without deception."[39] In the eleventh

O'Connor, following translations of Exodus 28:33, translates *mala punica* as pomegranate; Latham, *Revised Medieval Latin Word-List* (London: Oxford Univ. Pr. for the British Academy), gives the meaning orange (or lemon).

37.  P. E. Schramm et al., *Die deutschen Kaiser und Konige in Bildern ihrer Zeit, 751–1190*, 2nd ed. (Munich: Prestel Verlag, 1983), pp. 215–216; Patrick Corbet, *Les Saints Ottoniens: Sainteté dynastique, sainteté royale et sainteté féminine autour de l'an Mil* (Sigmaringen: Jan Thorbecke Verlag, 1986), p. 251, pl. 8; K. J. Leyser, *Rule and Conflict in an Early Medieval Society: Ottonian Saxony* (Bloomington: Indiana Univ. Pr., 1979), pp. 75–80, 101–102, pl. 4.

38.  O'Connor, "Star Mantle," pp. 75–76, 81–82, 101–102.

39.  Chrétien de Troyes, *Erec et Enide*, 6687–6747; L. T. Topsfield, *Chrétien de Troyes: A Study of the Arthurian Romances* (Cambridge: Cambridge Univ. Pr., 1981), pp. 57–63. Chrétien's account was brought to my attention by Tom Hill, E-mail of 27 June 1991.

and twelfth centuries astrology had found its place in court, but what was its status in the Carolingian age?

## The reemergence of astrology

Along with courtly displays of celestial symbolism, astrology, with its claims of celestial foresight, began to find a place in court culture. Emperor Louis the Pious was nearing his sixtieth year when he asked a member of his court, whom historians traditionally call "the Astronomer," the import of the comet that appeared in the evening sky. He expected to hear of the widespread belief that comets foretell the death of princes. The Astronomer, like many courtiers before and since, tried to turn his patron away from such concerns by quoting the prophet Jeremias that one should not fear signs in the heavens. Louis gently chided him for his reluctance to mention the ominous significance of comets.[40] This little story reveals much about astronomical knowledge in a ninth-century court. The Astronomer was certainly worthy of the name; he knew more than the practical astronomy of ecclesiastical computus and monastic timekeeping, for he described the position of the comet in terms of the classical constellations and he distinguished between its observed motion and the expected motion of the planets. The emperor, meanwhile, was clearly familiar with astrological lore, which both he and the Astronomer interpreted in an orthodox Christian fashion that saw the stars as signs rather than causes. In this court of a Christian emperor we see, if in rudimentary form, a reemergence of that traditional combination of astronomy and astrology which had flourished in the Roman Empire but had left scarcely a trace in the historical record of the early Middle Ages.[41]

Shortly after Louis's death in 842, Hrabanus Maurus wrote a treatise answering the questions of a certain Bonosus about marriages between cousins and about divination. The section *On the magical arts* condemns prognostication through communing with spirits more than it condemns the practice of astrology. *On the magical arts* does include Isidore's distinction of genethelialogical astrology, as named from its consideration of the position of the Sun and stars in the zodiac on the day of one's birth (*genesis*), from horoscopic astrology, as named from inferring a person's fate from the hour (*hora*) of one's birth. But both Isidore and Hrabanus used *horoscopus* in its general sense of determining the hour; they both seemed unaware of its technical meaning as that point on the zodiac that is rising

---

40. The Astronomer, *Vita Ludovici, PL* 104, col. 971–72.

41. Pliny, *Hist. nat.*, 2.22.89–2.23.93, describes comets, their significance, and their motion, which he says is like that of the planets. Hrabanus Maurus, *De computo*, 52.6–8, provides a contemporaneous statement that comets portend "the change of reigns, or pestilence, or war, or violent winds."

on the eastern horizon, a fundamental reference point used by astrologers in casting proper horoscopes.[42]

Hrabanus continued with a lengthy discussion, taken verbatim from Augustine's *De divinatione dæmonum*, of how demons can inform humans of future events. Augustine had taught that the subtle senses of demons can discern natural signs or causes from which they can, by drawing on their long experience, infer what is to come. But their prophecies are not certain and may be cunningly designed to deceive.[43] Since this is more a compilation than an original work, we cannot take Hrabanus's accounts here as accurately reflecting the divinatory practices of his time. He followed Isidore in distinguishing different kinds of astrology, but he also distinguished different aspects of the pagan Roman practice of casting auguries from the flight of birds. Some rudimentary forms of astrology may have been practiced in the ninth century, but Roman paganism was safely dead.

Yet Bonosus's inquiry about divination makes it clear that some form of divination was a current concern. Hrabanus's reply points to kings and courtiers as central to these practices. Hrabanus described, in passages that appear to be his own composition, the fates of Old Testament kings who had trusted in diviners: Saul's defeat after the witch of Endor called forth the spirit of Samuel (I Kings 28) and Achab's death after accepting promises of victory from false prophets (III Kings 22).[44] The role of kings in these examples suggests that divination at court troubled Hrabanus.

Hrabanus's concern with divination continues in a penitential that he wrote for Otgar, whom he succeeded as archbishop of Mainz in 847. Hrabanus told Otgar that he had compiled the penitential from the decrees of early church councils and synods to help him resolve difficult questions. Magic remained an important concern; Hrabanus sent a copy of his treatise *On the magical arts* along with the penitential.[45]

Hrabanus quoted a few decrees against astrology in the section on divination, including the prohibition against Christians following the traditions of the gentiles and noting the course of the Moon or stars. This prohibition is repeated, emphasizing that these meaningless signs are not to be used to determine when to build a house, or to plant grain or trees, or to consort with one's spouse. Hrabanus's presentation of these unqualified prohibitions on watching the stars, without comment, comes as a surprise, since he was a noted teacher of astronomy who had earlier adopted Bede's defense of the legitimate uses of the signs in the heavens in his scriptural commentaries. The reasonable inference is that Hra-

---

42.  Hrabanus Maurus, *De magicis artibus*, PL 110, col. 1097–99; Isidore of Seville, *Etymologiae*, 8.9.3–5, 9–35.

43.  Hrabanus Maurus, *De magicis artibus*, PL 110, col. 1101–07; Augustine, *De Divinatione dæmonum*, 3.7–10.14.

44.  Hrabanus Maurus, *De magicis artibus*, PL 110, col. 1100–01, 1108.

45.  John McCulloh, ed., Rabanus Maurus, *Martyrologium*, pp. xi–xxiv; Hrabanus Maurus, *Poenitentium liber*, Praef., PL 112, col. 1397–1399.

banus's repetition of stereotyped condemnations of astrology forms a minor part of a more general condemnation of divination and other pagan practices.[46]

Hrabanus did use his knowledge of astronomy when he admonished his congregation for their pagan reactions to an eclipse of the Moon. He tells how around sunset he was disturbed by the clamor raised by his neighbors, some sounding horns and some shooting arrows at the Moon to prevent monsters from devouring it, others fearing the growing darkness as a dreadful portent. He berated them for their lack of faith, reassured them of the natural causes of eclipses, and reminded them that if signs appear in the Sun or Moon or stars, it would be before the day of judgement.[47]

It is especially through rare events like Louis the Pious's encounter with Halley's comet or Hrabanus's reprimand of his raucous neighbors for lapsing into idolatry that we catch glimpses of general beliefs about celestial portents. Just as Louis the Pious and his astronomer were led to learned discussion to resolve the significance of comets to kings and commoners, so also were the common folk led to age-old rites to forestall the ominous darkness.

Although we can see popular beliefs in reactions to dramatic celestial events like comets and eclipses and in the spread of divinatory techniques that allude to the Moon and planets, we have little evidence that the practice of mathematical astrology, based on the computed positions of the heavenly bodies, was behind the ninth-century concern with divination. Hrabanus, despite his mastery of computus and concern with divination, gives no sign that he knew any more about horoscopic astrology than what he had read in Isidore of Seville. Furthermore, Hrabanus did not specifically attack the validity of astrology, although we have already seen the extensive body of Christian and pagan polemics against that art on which he could have drawn.[48] Finally, an absence of mathematical astrology is consistent with the limitations of ninth-century astronomy.

From what we know of the astronomical knowledge available to ninth-century astrologers, we can approach their art by asking what kinds of astrological computations they could make. The simplest form of genethelialogical astrology, which scarcely exceeds the "Are you a Scorpio?" kind of question, requires no mathematics beyond that found in a typical computus text. Finding the number of zodiacal signs separating the Moon from the Sun was a standard problem in computus since the time of Bede, but computus texts did not discuss how to find the place of the Sun in the zodiac. Although the problem was not treated, the essential data were there. The Sun's mean speed through the zodiac, conventional dates of the solstices and equinoxes, and sometimes the dates of the Sun's entry

46. Hrabanus Maurus, *Poenitentium liber*, 23–24, *PL* 112, col. 1397–1424; *De clericorum institutione*, 25, *PL* 107, col. 403–404.
47. Hrabanus Maurus, *Homilia*, 42, *PL* 110, col. 78–80.
48. Hrabanus Maurus, *De magicis artibus*, *PL* 110, col. 1097–99.

into each sign of the zodiac were common elements in most computi. With these
an astrologer could compute the approximate positions of the Sun and Moon in
the zodiac for any day of the year.[49]

Computing the past or future places of the five planets is more complex. Yet
the computists gave only the average period in which each planet passes around
the zodiac, rounded to the nearest year and with no indication of when the planets
begin their circuits through the signs. Hrabanus's de computo mentions the planets'
retrograde motion, but only qualitatively and with no hint of the mathematics
required to compute how retrograde motion would modify the planets' mean
positions.[50] At one point Hrabanus discussed the current place of all seven planets
in the zodiac; he computed the positions of the Sun and Moon to the nearest
part (Hrabanus's part was approximately a degree, as he divided the zodiac into
365 1/4 parts). He located Mars, Jupiter, and Saturn within a sign, apparently by
observation since he noted that Venus and Mercury were too close to the Sun
to determine their signs. Computus did not provide an adequate theoretical basis
for locating the five planets in the zodiac.[51]

Horoscopic astrology introduces further complexities; the astrologer must be
able to place the Sun, Moon, and other planets in their proper astrological
"houses" related to the horizon and mid-heaven.[52] This element of horoscopic
astrology concerns the rising of the zodiacal signs; for example, which signs are
rising, which setting, which in mid-heaven, and which at the nadir at any given
time of the day and year. Mastery of these aspects of astrology would require a
spherical astronomy significantly beyond the astronomy and geometry taught in
monastic schools.[53]

Since the mathematical basis of horoscopic astrology was not available, we must
ask what, if any, kind of astrology was practiced in the ninth century. Ninth-
century Latin divinatory texts indicate the extremely rudimentary kinds of astral

49. Bede, De temporum ratione, 17.5–17, 16.33–58; Rabanus Maurus, De computo, 40.10–13; 53.6–11,
    48–55.

50. Bede, De temporum ratione, 8.42–48; Rabanus Maurus, De computo, 38; 37.39–45.

51. Rabanus Maurus, De computo, 48.14–19. His positions of Mars, Jupiter, and Saturn place them
    within the same sign as do modern calculations. His positions of the Sun and Moon are not as
    accurate as he claims, reflecting errors in computistical calculations of the equinoxes. Tuckerman,
    Planetary Positions, vol. 59.
        Pingree suggests that Hrabanus computed these positions using the Preceptum canonis Ptolomei.
    Yet no extant manuscript of the Preceptum contains tables for the five planets, and Hrabanus's
    computus does not mention the Egyptian year of 365 days used in the Preceptum. Pingree,
    "Preceptum canonis Ptolomei," pp. 371–372.

52. Those interested in the mathematical complexities of astrology and its different traditions should
    consult North, Horoscopes and History.

53. Some simpler astrological systems avoided the complexities of spherical geometry, but even these
    simpler schemes were not known in the Carolingian schools.

divination that emerged during the Carolingian Renaissance.[54] Manuscripts from the monasteries of Laon and Corbie provide a Latin translation of a Hellenistic text attributed to a pagan Egyptian priest, Petosiris. This text describes in a few pages a Greek method of divination based on summing the numerical value of the letters of the supplicant's name (transliterated into Greek) and a number assigned to the age of the Moon on the crucial date for which the supplicant's question is to be resolved. After dividing the sum by twenty-nine, the remainder is used to read the outcome from a circular diagram. An auxiliary table gives the numerical values of the Greek names of the planetary gods, arranged in the order of the days of the week. This provides a simpler means of prognostication from the day of the week if the soothsayer did not know how to compute the age of the Moon.[55]

Another brief translation from the Greek, the so-called Revelation of Esdras, documents year-end prognostications that were also hinted at in Hrabanus's penitential.[56] This brief text, which circulated in Eastern Christendom as early as the seventh century, gives simple rules to foretell the course of the coming year from the day of the week on which the Kalends of January fell, the day on which the year begins. The earliest Latin manuscripts date to the ninth century, from the Carolingian monasteries of Fleury and Lorsch, and the text soon spread widely in Latin and the vernacular languages.[57] The rapid spread of these texts suggests a widespread fascination with astral prognostication which would offer a fertile ground for the later growth of a geometrically based horoscopic astrology.

## Teaching computus

A series of new textbooks composed to teach the arithmetical astronomy of computus lay closer to the heart of the Carolingian educational reform than did pop-

54. Thorndike, History of Magic, vol. 1, pp. 672–696; van de Vyver, "Les plus anciennes Traductions."

55. Thomas G. Tolles, "The Latin Tradition of the Epistola Petosiridis," Manuscripta, 26(1982):50–60; van de Vyver, "Les plus anciennes Traductions," pp. 674–676.

56. Hrabanus quotes a condemnation of celebrating the Kalends by resting from work or decorating one's house with garlands of greenery. Here he is again attacking a group of ancient customs, these associated with the Roman festival on the Kalends of January at which the course of the coming year was foretold. Hrabanus Maurus, Poenitentium liber, 24, PL 112, col. 1417; Michel Meslin, La Fête des kalends de janvier dans l'empire romain, Collection Latomus 115 (Brussels: Revue d'Études Latines, 1970), pp. 13, 24, 105–107.

57. E. Ann Matter, "The 'Revelatio Esdrae' in Latin and English Traditions," Revue Bénédictine, 92(1982):376–392. Dr. Matter kindly brought this text to my attention, E-mail of 23 May 1991. On later influence, see Max Förster, "Beiträge zur mittelalterliche Volkskunde, 11. Mittelenglische Bauernpraktiken," Archiv für das Studium der neueren Sprachen und Literaturen, 128(1912): 291–297.

ular divinatory texts or lavish astronomical compilations. In the summer of 820 Hrabanus Maurus, master of the monastic school at Fulda, wrote an influential computistical text.[58] Although Hrabanus's content is traditional, his style reflects dialogues between master and disciple like those his own master, Alcuin, had used in his teaching at Aachen and Tours:

D.  From whence come the names [of the days of the week] that moderns use?

M.  From antiquity, that is, from the superstitions of the peoples. . . .

| D. | How many hours are there in a week? | M. | 168. |
|----|-------------------------------------|----|------|
| D. | How many points? | M. | 662. |
| D. | How many minutes? | M. | 1,680. |
| D. | How many parts? | M. | 2,520. |
| D. | How many moments? | M. | 6,720. |
| D. | How many ostenta? | M. | 10,080. |
| D. | How many atoms? | M. | 3,790,800.[59] |

These calculations of ever smaller units of time seem to have been little more than students' exercises, with little real application. One wonders how seriously these minutiae were to be taken, considering that the number of points should have been 672, the number of atoms 3,790,080. But Hrabanus did not restrict himself to the traditional content of the computus; he also included material from the new Carolingian texts. His students were interested in the constellations as well as the Sun, Moon, and other planets.

D.  Please explain to me, as you had promised, the position of the stars.

M.  There are (as Aratus tells in his *Phenomena*) thirty signs besides those which make up the zodiac, some of which rise in the north and others which are set off to the south. . . . [60]

Hrabanus's discussion is based on the established authorities, but he goes beyond Bede to include such a newly rediscovered authority as Aratus. Unlike Bede and Alcuin, however, Hrabanus shows little of their feel for astronomical observations.

Both new content and a different teaching style appear in the introduction to computus written, apparently by Helperic of Auxerre, near the end of the ninth century. Helperic admitted that his text was simple, aimed at providing his young

---

58.  Wesley Stevens, ed., Rabanus Maurus, *De computo*, Introduction, pp. 176–189. On Hrabanus's influence on his student Walahfrid Strabo, later abbot of Reichenau, see Stevens, "*Compotistica et astronomica*."

59.  Rabanus Maurus, *De computo*, 26.12–14, 27.14–21.

60.  Rabanus Maurus, *De computo*, 51.2–6.

students with the ABCs of computus in simple language. Yet, like his predecessors, he sometimes digressed from the practical problems of the calendar to illuminate computus from unusual directions. Perhaps these merits account for the wide and lasting influence of his text, which was used for over two centuries and became part of the standard corpus of computistical texts.[61]

Helperic's teaching reflected a new awareness of the rudimentary astronomy of Roman astronomical writers, but he did not slavishly follow ancient authorities. Helperic cited Macrobius's commentary on Cicero's *Dream of Scipio*, recommending it to those students who wished to study astronomy in greater detail. Yet he argued against those who maintained the standard astronomical view that the Sun and the other planets move against the general motion of the universe. Helperic pointed out the flaws in Macrobius's unconvincing argument for this direction of motion, but his only reason for preferring the alternative view was that it seemed "simpler to him" for the Sun to have one motion.[62]

Helperic also presented his students with a simple empirical method to confirm the traditional dates of the solstices and equinoxes. His technique is reminiscent of Bede's discussions of observing the position of the equinoctial Sun on the horizon and of earlier traditional methods for observing the Sun's motion qualitatively and without instruments. Like Bede he described the day-to-day change of the place of sunrise noting that the rising Sun becomes "higher" as its rising point moves northward along the horizon from winter to summer and becomes "lower" as it moves towards winter.[63]

Helperic went beyond Bede, however, to recommend that his reader should note each day at sunrise where sunbeams, passing through an aperture, fall on the western wall of a room.[64] Helperic claimed that this would demonstrate that the Sun first turns back at the winter solstice on 22 December and that between the eighteenth and the twenty-first the Sun will be moving down towards the south, while its ray on the western wall will be moving towards the north. He also proposed a similar observation as the Sun completes its ascent at the summer solstice, which falls 182 days later, on 21 June. Turning to the traditional computistical emphasis on counting days, he noted that on leap years the additional day is added during the ascent of the Sun, making the period of ascent from winter solstice to summer solstice equal to the 183 days of descent from summer

61.  Helperic's text spread widely and was used into the middle of the twelfth century. Some eighty copies survive with the current year for sample calculations ranging from 900 to 1151. Traube, "Computus Helperici"; McGurk, "Computus Helperici"; van de Vyver, "Oeuvres d'Abbon de Fleury," pp. 147–149; Jones, *BOT*, p. 130, n.2; Helperic, *Liber de computo*, 23, *PL* 137, col. 37.

62.  Helperic, *Liber de computo*, 2.4, *PL* 137, col. 23–27; Macrobius, *In Somnium Scipionis*, 1.18.

63.  "Higher" and "lower" cannot refer here to height above the horizon, nor to the changing height of the noontime Sun. Most likely "height" is observed in reference to the slanting path of the rising or setting Sun. Cf. Bede, *De temporibus liber*, 10.12–18; *De temporum ratione*, 38.40–48; letter of Ceolfrith of Jarrow to Nechtan, in Bede, *Hist. eccl.*, 5.21.

64.  Helperic, *Liber de computo*, 31, *PL* 137, col. 40–41.

solstice to winter solstice. At the equinoxes, which he placed midway in time between the solstices, the ray will shine due west and fall between these two extreme positions on the western wall.[65]

Helperic's observations cannot be described as exact, for at his time the winter solstice fell on 16 or 17 December, the summer solstice on 17 June. Yet this observational error is not unreasonable; at the solstices the Sun scarcely moves from day to day, and it is very difficult to mark the specific day it pauses at the solstices. Like many other watchers of sunrise, he could observe the place where the Sun pauses at the solstices, even though he would not be sure exactly when it got there. But Helperic was not trying to make a scientific measurement; he wanted to demonstrate to his students how the solstices divide the year. Given the inherent lack of precision of these observations, their ambiguous results could be read as confirming his, and his students', expectations.

Abbo, of the monastery of Fleury, continued the dissemination of computus to a wider audience and extended the scope of the discipline in an attempt to apply it to the motion of the five planets. Around 978 Abbo wrote four short astronomical treatises for his students.[66] Three of these treatises deal with the celestial spheres, describing the celestial spheres, the circles of the world, and the motions of the planets through the zodiac. A fourth treatise discusses the heliacal and cosmical (or morning and evening) risings and settings of the stars, to which Abbo gives the proper Greek names ἡλιακός and κοσμικός.

In his De ratione spere, a treatise falling somewhere between the computistical anthologies of the Carolingian schools and the more complex treatises on the sphere that would appear in the universities of the thirteenth century, Abbo presented not only the celestial sphere and its circles but also the motion of the planets. Following the Plinian tradition, Abbo discussed the planets' retrograde motions and noted that Mercury and Venus are never more than one or two signs from the Sun. Like Macrobius, he reported erroneously that each sign of the zodiac rises in exactly two hours. His truly original contribution was an attempt to compute the positions of the five planets on the assumption that they moved uniformly through the zodiac from the time of Creation. Abbo took as his starting point Macrobius's statement that the planets were created in their apsides, that is, with Mars in Scorpio, Jupiter in Sagittarius, Saturn in Capricorn, Venus in Libra, and Mercury in Virgo. He then assumed that the five planets moved uniformly with periods of 6, 12, 18, 24, and 30 years.[67]

65. Helperic, Liber de computo, 31, PL 137, col. 40–43.
66. Abbo of Fleury, "Two Astronomical Tractates"; "Further Astronomical Material." On Abbo's sources, see Wallis, "MS St. John's College 17," pp. 391, 405–406; Borst, Plinius und seine Leser, pp. 212–213.
67. Abbo of Fleury, De ratione spere, 1–67, 170–185; Macrobius, In Somnium Scipionis, 1.21.24. Macrobius's apsides differ from those of Pliny, Hist. nat., 2.13.64–65.

The results of Abbo's proposed computations could never have survived any comparison with observations of the planets. Clearly the periods are based on numerological principles rather than observational data. Abbo's assumption of uniform motion made no attempt to account for the planets' retrograde motions, which he had already described. Since Mercury was given a period of twelve years to circle the zodiac, and Venus twenty-four, while the Sun makes its circuit in only one year, these calculations would often find the two inferior planets opposite the Sun, although Abbo had noted that they were never far from it. In particular, Abbo had Venus in Libra and Mercury in Virgo at the moment of Creation, directly across the zodiac from Aries, where the computists placed the Sun on the first day.[68]

Abbo's attempt to extend computus to deal with the problem of the planets reflects an active curiosity about astronomical problems, but his simple arithmetical approach was inadequate to deal with even the most obvious elements of planetary motion. His knowledge of spherical geometry imposed similar limits. Although he could describe the circles of the celestial sphere, he was unaware that the inclination of the zodiac caused each sign to take a different time to rise. In his commentary on the *Calculus* of Victorius of Aquitaine, he followed Macrobius's view that each sign rose in exactly two hours.[69]

In his computus Abbo continued the simple teaching of the monastic school. Where Bede had proposed a simple table using letters of the alphabet to indicate the phases of the Moon for those unable to master the complexities of simple arithmetic, Abbo went even further to meet the limits of his students. Abbo's computus incorporated Bede's table, the "Dominical" or Sunday letters indicating the days of the week following the practice of ancient Roman calendars, and other tables from which one could read other computistical parameters without calculation.[70]

In 986 Abbo was called to England to teach for two years at the monastery of Ramsey. This was a time of a major revival of English monastic life after the devastation of centuries of Viking invasions. Important astronomical and computistical manuscripts, including copies of Abbo's works and of the Carolingian computistical collection, were brought from Fleury to England. These manu-

---

68. Macrobius had placed the Sun in Leo at the Creation, near Mercury and Venus; Abbo did not discuss the Sun's place at Creation. Macrobius, *In Somnium Scipionis*, 1.21.24; Abbo of Fleury, *De ratione spere*, 53–65.

69. Evans and Peden, "Natural Science in Abbo of Fleury." Dr. Peden graciously provided me with a copy of her transcription of this passage from Berlin, MS Phillipps, 1833, fol. 13v. Cf. Macrobius, *In Somnium Scipionis*, 1.21.11–22.

70. A. Cordoliani, "Les Manuscrits de la Bibliothèque de Berne Provenant de l'Abbaye de Fleury au XIᵉ Siècle: Le Comput d'Abbon," *Zeitschrift für schweizerische Kirchengeschichte*, 49(1955):135–150; Jones, *Bedae Pseudepigrapha*, p. 61.

scripts were the font of a series of English copies, while Abbo's teaching produced a number of disciples, the most noteworthy being Byrhtferth of Ramsey.[71]

Byrhtferth wrote his *Manual* largely in the vernacular, continuing the tradition of an Anglo-Saxon literary culture through which what was "clearly understood by monks . . . [would also] be known by clerks."[72] Byrhtferth placed computus, or *rimcræft*, at the heart of a broader study of the natural world. While the basics of his *Manual* are taken from Bede and Hrabanus, Byrhtferth drew upon other areas of the quadrivium to discuss the natural and mystical significance of numbers in the universe, which God had established "in measure, in number, and in weight."[73]

Marking the culmination of this computistical tradition is a manuscript written a century later only a few miles from Ramsey. It goes beyond the Carolingian computistical anthologies to include the work of Abbo, Byrhtferth, and Helperic, as well as that of such earlier authorities as Bede, Isidore, and Pliny, while omitting the "star catalogs" found in the Carolingian texts. This visually striking manuscript is known for its many diagrams that illustrate a wide range of astronomical topics. One circular diagram (or *rota*) is placed among three other *rotae* that show the possible age of the moon on Easter under the reckonings of Dionysius, Victorius, and "Antiochos." This fourth *rota* is a T-O map, a common schematic representation of the locations of Asia, Europe, and Africa. Rather than marking the directions with the twelve classical winds, this map uses the places of sunrise and sunset at the birth of St. John (the summer solstice), at the equinoxes, and at the birth of the Lord (the winter solstice), as well as the directions of the Sun at midday and midnight (Fig. 23). Like the other *rotae*, it deals with the liturgical calendar, but it does this by relating the calendar's temporal framework to the spatial framework of a solar horizon calendar.[74]

The solstices and equinoxes reappear in this manuscript's most complex illustration, Byrhtferth's cosmological diagram (Fig. 24). The diagram illustrates how

---

71.  Van de Vyver, "Oeuvres d'Abbon de Fleury," pp. 143–144, 163–164; Evans, "Abacus in English schools," pp. 72–74; Wilhelm Koehler and Florentine Mütherich, *Die Hofschule Kaiser Lothars: Einzelhandschriften aus Lotharingien*, Karolingischen Miniaturen, 4 (Berlin: Deutscher Verlag für Kunstwissenschaft, 1971), pp. 77–79.

72.  *Byrhtferth's Manual*, 10.26–7.

73.  *Byrhtferth's Manual*, 8.16–18, quoting Wisdom 11:21.

74.  Oxford, St. John's College, MS 17, fol. 5v; Wallis, "MS St. John's College 17," pp. 216–219. Directions are often marked with the twelve classical winds on T-O maps associated with Isidore's *De natura rerum* and Bede's *De temporum ratione*. Destombes, *Mappemondes*, pp. 29–36; Otto Homburger, *Die illustrierten Handschriften der Burgerbibliothek Bern: Die vorkarolingischen und karolingischen Handschriften* (Bern, 1962), figs. 13, 74; Obrist, "Diagramme isidorien," figs. 16, 17, 30; Obrist, "Wind Diagrams," figs. 9, 10, 12, 13, 18.

   Another figure on fol. 35v ambiguously depicts the places of sunrise and sunset at the solstices and equinoxes; it most probably represents the diurnal paths taken by the Sun from its rising points on the eastern horizon, through the meridian, and back to the western horizon. A similar figure (Cologne, Dombibliothek, cod. 83[II], fol. 81v) continues these paths below the horizon.

Figure 23. T-O map and Paschal rota. At the center of the lower circle is a T-O map depicting Asia, Eruope, and Africa. The directions are marked by places of the sun, *reading clockwise from the ten o'clock position:* "Sunrise on the birth of St. John. Sunrise on the equinox. Sunrise on the birth of the Lord. Midday. Sunset on the birth of the Lord. . . ." Oxford, St. John's College, MS 17, fol. 5v. Used by permission of the President and Scholars of St. John's College, Oxford.

the number four ties together the four elements, earth, water, air, and fire; the four sensible qualities, hot, cold, wet, and dry; the four seasons, here divided by the conventional dates of the two solstices and the two equinoxes; the four ages of man; the four cardinal directions, East = *Anatole*, West = *Disis*, North =

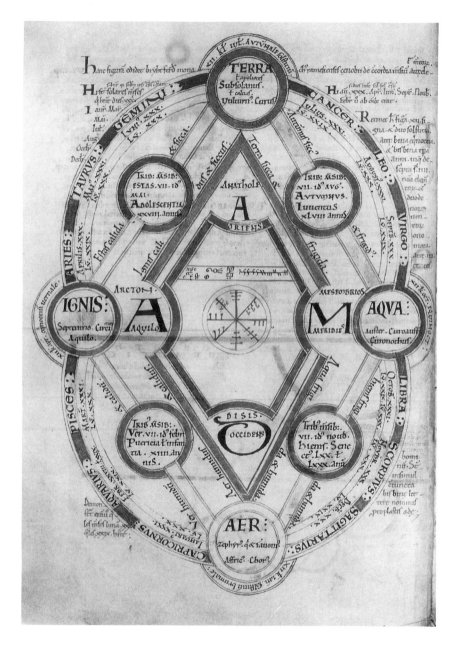

Figure 24. Cosmic symbolism of the number four. Oxford, St. John's College, MS 17, fol. 7v. Used by permission of the President and Scholars of St. John's College, Oxford.

*Arcton*, and South = *Mesembrios*; and the four letters of the name of the first-formed man, ADAM.[75] In the computus that Abbo and Byrhtferth and their colleagues taught in monastic schools, number was at the core of the understanding of nature.

## The revival of the liberal arts

Along with its relation to the evolving discipline of computus, astronomy found another place in the curriculum as the works in which the liberal arts had been codified at their twilight in late antiquity were studied and taught. During the ninth century scholars increasingly commented on ancient astronomical ideas as a routine part of their teaching of the liberal arts.

The many roles of astronomy at the court of Charles the Bald were epitomized by the poet, philosopher, and teacher John Scottus Eriugena (ca. 810–ca. 877). In his poem the "Aulae siderae," written for the dedication of the palace church, Eriugena drew upon the traditional bond associating rulers with astronomy, filling his poem with astronomical metaphors and numerical harmonies and their spiritual significance. He spoke of the solstices and equinoxes and their relation to the births of Christ and John the Baptist, the circle of the zodiac and the Sun that binds everything with its rays, and the stars, clad in a mantle illumined by the light which Christ brought to the world in its season of darkness. All these celestial symbols find their reflections in the palace church.[76]

But Eriugena is not best known for his poetry. His most significant achievement was the first attempt by a medieval scholar to write a self-standing treatise on philosophy, his *On the Divisions of Nature*, or *Periphyseon*. Although Eriugena's philosophy benefited from his knowledge of Greek, his discussions of astronomy did not go beyond those found in Latin astronomical texts.

Eriugena dealt with the heavens in Book Three of the *Periphyseon*, where he treated the creation of the material universe following the scriptural sequence of the six days of Creation. In his discussion of the material nature of the firmament, created on the second day, he alluded to the central place of the Sun, midway between the earth and the stars. He then asserted that "Jupiter and Mars, Venus and Mercury . . . always pursue their orbits around the Sun."[77] Historians have disagreed about whether Eriugena meant to claim that the Sun is the geometric center of these planets' orbits or merely that these planets differ from chilly Saturn

---

75. *Byrhtferth's Manual*, 200.10–204.18; Wallis, "MS St. John's College 17," pp. 230–238, 789–816. Wallis discerns many theological symbols in Byrhtferth's diagram, seeing it as an abstract "*majestas Domini* encoded in the symbolism of . . . the computists' time, centered on the date of Easter." P. 815.

76. O'Meara, *Eriugena*, pp. 178–188. O'Meara's discussion includes the text and an English translation.

77. Eriugena, *Periphyseon*, 3.698A.

in that they are near the Sun, are physically like it, and traverse the same region of the heavens that it does.[78] Whatever the intent of this ambiguous passage, Eriugena did not provide his contemporaries with either a geometrical or an observational basis for a heliocentric model.

When Eriugena reached the fourth day, in which the two great luminaries were created, he launched into extended discussions of the dimensions of the earth, Sun, and the Moon and the harmonious spacing of the spheres that carry the seven planets. These subjects had become current with the appearance of the Carolingian anthologies, and Eriugena tried to clarify and resolve inconsistencies among the traditional sources excerpted in those anthologies. Eriugena knew that these dimensions were based on observations, yet his geometrical skills were inadequate to define the dimensions of the celestial bodies. For example, he repeatedly gave the circumference of a circle as twice its diameter.[79] These passages indicate how misunderstandings of geometrical concepts or terminology still hampered the astronomy of the liberal arts, despite the increasing philosophical sophistication of the ninth century.

Equally influential in the integration of the liberal arts into the curriculum were textual commentaries reflecting the teaching of Eriugena and his successors, especially Remigius of Auxerre. Remigius had been master of the school at Auxerre from 876 to about 883, from which he was called to teach at Rheims. From around 900 until his death about 908 he taught at Paris, apparently in the monastic school of St. Germain-des-Prés. There he taught and wrote on many subjects, including ancient mythology, Boethius's *Consolation of Philosophy*, and Martianus Capella's *The Marriage of Philology and Mercury*.[80] It is clear that teaching was no longer limited to the excerpts from ancient learning found in the Carolingian anthologies, or Bede, or Isidore. Remigius taught his students from complete ancient texts, interpreting them within the context of Christian thought and his own familiarity with astronomy.

His commentary on Boethius reveals a distinct shift in orientation from the author of the *Consolation*. Remigius followed Boethius in maintaining that all creation is governed by reason; he commented extensively on the verse where Boethius had spoken of the Father of heaven and earth who perpetually guides the world by reason. But where Boethius had spoken philosophically, Remigius reflected the ambiguity of medieval approaches to divine reason. At one point he treated the divine reason in an Augustinian manner as the "forms, exemplars and reasons in the mind of God" after which the world is patterned; elsewhere he

78.  Duhem, *Le Système du Monde*, vol. 3, pp. 58–62; Erika von Erhardt-Siebold and Rudolf von Erhardt, *The Astronomy of Johannes Scotus Erigena* (Poughkeepsie, N.Y.: Vassar College, 1940).

79.  Eriugena, *Periphyseon*, 3.718A, 722D.

80.  Pierre Courcelle, *La Consolation de Philosophie dans la tradition littéraire: Antécédents et Postéritie de Boéce*, Études Augustiniennes (Paris: Études Augustiniennes, 1967), pp. 241–274; Remigius of Auxerre, *In Martianum Capellam*, vol. 1, pp. 1–15.

personified this reason as "the Son of God through whom all things were created and are governed." The former points towards an almost self-sufficient natural order, while the latter shifts from the rational order itself towards its ultimate dependence on the will of God.[81]

To the extent that Remigius was familiar with ancient philosophy, it is that part of Plato's *Timaeus* as expounded by Chalcidius which had focused on an explanation of the material world. Since Boethius's discussion with Lady Philosophy looked at earthly things from a celestial perspective, Remigius took the opportunity to introduce his bits of astronomical knowledge, sometimes even taking issue with Boethius. Boethius had wrongly attributed to Ptolemy the view that the earth is divided into four climates; Remigius, not having read Ptolemy, accepted Boethius's assertion, but he pointed out that according to Hyginus and the other *astrologi* the heavens, and the earth beneath the heavens, are divided into five regions.[82] At one point, in discussing the relation of fate and divine providence, Boethius had mentioned the possibility that fate may take effect through the motions of the stars. Remigius notes that some (*quidam*) had considered Boethius as coming close to heresy for this view, but Remigius exonerated Boethius from any suspicion of astral causation; Boethius was merely enumerating the various possibilities, without stating his own position.[83]

If Boethius provided Remigius and his successors with an opportunity to introduce their students in broad terms to the order of the universe, Martianus provided an ideal text in the liberal arts – and especially in astronomy – the discipline that later commentators would associate with him.[84] The allegorical structure, technical terminology, and challenging content of the *De nuptiis* offered abundant opportunities for comment, definition, and explanation on everything from grammar and mythology to mathematics and astronomy.

Remigius's commentary on Martianus Capella relied so heavily on Eriugena's that they are best treated together. Like Eriugena, Remigius opened his discussion of *astrologia* or *astronomia* with Isidore of Seville's definitions: astrology concerns the reason or account (*logos*) of the stars, while astronomy concerns the laws (*nomos*) that govern the stars' risings, which vary with one's position on the earth. Neither of these sciences included astrological divination.[85]

Given the mythological context of Martianus's work, Astronomy appears

---

81. Remigius, *In Boethius*, ed. Silk, pp. 333–334; ed. Stewart, pp. 30–31; ed. Silvestre, pp. 51–53.

82. Boethius, *De consolatione philosophiae*, 2, pr. 7; Remigius, *In Boethius*, ed. Silk, p. 326.

83. Remigius, *In Boethius*, ed. Stewart, p. 37.

84. *Causa efficiens [astrologiae]: Martianus.* Anonymous, *Accessus philosophorum .vii. artium liberalium*, 645, in Lafleur, *Quatre introductions*, p. 219. "Illa autem que dicitur demonstrativa vel demonstrabilis dividitur in duas, quia quedam . . . ostenditur passionem de subiecto per causam, et illa traditur in *Almagesti*; altera autem, que docet quid est quod dicitur per nomen, et illa traditur in ipso Martiano." Anonymous, *Compendium circa quadrivium*, 91–97, in Lafleur, p. 363.

85. Remigius, *In Martianum Capellam*, 422; Eriugena, *In Marcianum*, 422; cf. Isidore of Seville, *Etymologiae*, 3.24–27. Eastwood, "Astronomies in the Carolingian World."

among the planetary gods "who determine men's destinies." Remigius went beyond Eriugena to caution his students against the "fictions of the mathematicians" who would determine events from the stars, against the pagan view that the constellations are causes, and against Martianus's gods of air, earth, and sea that are accepted by rustics and common folk who do not know *astrologia*'s reasoned account of the stars' motion. Remigius seems to have been concerned that even his students may have been influenced by folk practices; he repeatedly reminded them that Martianus used the word *auspicium* to mean beginning, directing them away from the more common sense of augury.[86] Despite his concern with astrological prognostications, Remigius seemed unaware of the astrological sense of the word *horoscopus*. He used it with no misgivings, along with *episcopus*, bishop (literally, overseer), to clarify for his students the meaning of the term *scopus*, philosophical speculations or observations. Like Eriugena, Remigius considered a horoscope as concerned with observing the hours, with timekeeping; hence he glossed Martianus's obscure term *clepsydra*, or water clock, as a horoscopic vase.[87]

Besides showing a decided lack of familiarity with technical astrology, these exercises with the meaning of words suggest that Eriugena and Remigius were more concerned with a literary study of the text than with a geometrical investigation of the heavens. When addressing Martianus's discussion of the nonuniform rising of the signs of the zodiac, neither author gave a reason for that nonuniformity. Remigius, however, went beyond Eriugena's gloss on Martianus's fractional units of time to employ various fractional parts of an hour, a topic the computists had long enjoyed expounding.[88]

Remigius also knew enough astronomy to correct some minor points in the text that Eriugena ignored, and the astronomy he used was the astronomy of computus. He knew the time it took the Moon to circle the zodiac, charitably reinterpreting the text's erroneous value of 27 2/3 days to the correct value of 27 days and 8 hours. Where Martianus had said that the longest day of the year was 14 1/6 hours, Remigius noted that Martianus was describing his own climate, but at "our own" climate the longest day is 18 hours long. This value does not appear in his antique sources; it comes from the tradition of ecclesiastical computus, appearing in one of the calendars associated with the computistical anthology of 809.[89]

The most extensively glossed portion of Martianus's text treats the dimensions of the celestial spheres. This reflects the influence of Carolingian astronomy, for

86. Martianus Capella, *De nuptiis*, 8.810; Remigius, *In Martianum Capellam*, 428.14–18, 294.3, 437.10; Eriugena, *In Marcianum*, 294.3.

87. Remigius, *In Martianum Capellam*, 429.18, 446.11, 452.13, 295.4–5; Eriugena, *In Marcianum*, 295.4–5.

88. Remigius, *In Martianum Capellam*, 444.7–10; Eriugena, *In Marcianum*, 444.9–13; Rabanus Maurus, *De computo*, 27.14–21.

89. Remigius, *In Martianum Capellam*, 445.23, 446.3. See the entry for the longest day (19 June) in McCulloh, "*Martyrologium Excarpsatum*."

these passages had been excerpted in the Carolingian seven-book computus and the problem had been discussed in Eriugena's *Periphyseon*. This broad cosmographical question of the size and order of the celestial spheres had become more important than determining the courses of the stars and their risings and settings.

Of course, at this time when various astronomical strands were drawing together, Remigius knew a smattering of ancient astronomy as well. His commentary occasionally explains Greek terms, some taken from Martianus, some drawn from other sources, few fully understood. Drawing on Martianus's view that Venus and Mercury, unlike the other planets, proceed on "epicycles" that do not encompass the globe of the earth, Remigius improved on Eriugena's definition of epicycles, which had described them as "circles suspended above," to specify that they were either above a circle or above the circle of the earth and pass above the earth. Where Martianus had followed Pliny, listing the apsides of each planet, where they are farthest from the earth, Remigius believed that there were two apsides, one where the planet is farthest from the earth and the other where it is closest. As to the meaning of the term, he accepted Pliny's definition of *apsis* as the Greek word for circle, while Eriugena reckoned it was a circle elevated (in what way he does not tell) above another circle.[90]

Remigius often sought to interpret Greek astronomical terms, which Eriugena prudently avoided. His confusion reached its apex when he introduced that obscure technical term, *anabibazontem*, or ascending node, which we have already seen in the Latin translation of Ptolemy's *Canones* and Chalcidius's commentary on Plato's *Timaeus*. Remigius translated it as "climbing upwards," but sometimes he used it as a substitute for "meridian," although it is not quite clear whether he knew what the meridian is. More frequently he took it to mean the line that the Sun follows and at which eclipses can take place, that is, the ecliptic. Once both he and Eriugena used it as an adjective, this time in Greek letters *ANABIBAZONTA*, to refer to an "ascending conjunction" of the Sun and Moon, as opposed to a "descending conjunction," or *KATABIBACIN*.[91]

To the extent that Remigius's teaching on the liberal arts reflects the state of astronomical study in the middle of the ninth century, we can see a change just beginning to take place. Remigius, like most of his contemporaries, knew the computus, but the texts on the liberal arts he was discussing were clearly different. Yet he felt sufficiently confident to draw on his knowledge of computus to clarify, and even correct, Martianus's text. His occasional use of Greek technical terms

---

90. Remigius, *In Martianum Capellam*, 464.4, 467.12; Eriugena, *In Marcianum*, 464.5, 467.12.

91. Remigius, *In Martianum Capellam*, 298.15, 432.15, 457.21, 458.1, 458.17, 459.10–14; cf. Eriugena, *In Marcianum*, 459.15. The use of *anabibazon* to refer to a line is reminiscent of Chalcidius's commentary on the *Timaeus*, where it is one end of the same line that is a diameter of both the circle bearing the Sun and the circle bearing the Moon; Chalcidius, *In Timaeum*, 139.13–18. Remigius's statement that the opposition or conjunction must be on this line if the Moon or Sun is to be eclipsed recalls a number of passages in the *Praeceptum canones Ptolemei*, MS BL, Harl. 2506, ff. 60v, 61r.

indicates that he knew some ancient sources, probably through antique translations and commentaries, although we cannot rule out that astronomical texts were
among that feeble trickle of Greek texts arriving in the Carolingian court. Yet
when he drew on ancient sources to venture into the realm of geometrical astronomy, Remigius's grasp of the subject became increasingly insecure. It was
computus, not ancient learning, that provided Remigius with his key to astronomy.

Further changes to the astronomy of the liberal arts appear in an anonymous
cosmological treatise, *On the Universe and the Soul*, which may have been written
as early as the ninth century but was probably written in the eleventh. The text
gives no sign of the new astronomy received from the Islamic world and only
the slightest hints of the Aristotelian philosophy that would later dominate the
medieval universities; the topics reflect the scope and limits of astronomical teaching before the translations of the twelfth century.[92] It includes traditional discussions of the various circles of the heavens and the relations of the epicycles of
Venus and Mercury to the Sun. But our author adds discussions of the astrological
influences of the planets and their domiciles in the zodiac.

More significant for astronomy is an original discussion of a traditional topic,
the motion of the Sun and other planets through the zodiac. The author elaborates
on the computists' concern with the solstices and the equinoxes, and his method
reflects the debt to the computists that he shares with his contemporaries.[93] Unlike
the computists, who had treated the Sun's motion through the zodiac as if it were
uniform, our author sought to explain the classical accounts of the Sun's nonuniform motion through the zodiac. He began with the computists' conclusion
that the Sun passes through each sign in 30 days, 10 1/2 hours, found by dividing
the length of the year (365 days and 6 hours) by twelve.[94] He then noted that
certain unnamed philosophers, clearly including Martianus Capella, knew that the
Sun takes 2 more days than this average to pass through Gemini and 2 days less
to pass through Sagittarius.[95] Although he related this to the Sun's greater distance
from the earth in Gemini, he did not use geometry to compute the changing
time the Sun takes to pass through each sign. Instead, he used a simple linear
scheme in which the Sun took 16 hours more to pass through each successive

92. Ps.-Bede, *De mundi constitutione*.
93. Ps.-Bede, *De mundi constitutione*, 311–327.
94. For example Bede, *De temporum ratione*, 16.27–30; Rabanus, *De computo*, 40.10–11.
95. Martianus had noted that the Sun passes through the sign of Gemini in 32 days and through
    Sagittarius in 28 days, and that in the intervening signs you either add or subtract. Remigius of
    Auxerre and others made the same point in their commentaries on Martianus. None of them,
    however, incorporated the computists' value found in *De mundi constitutione*. Martianus Capella,
    *De nuptiis*, 8.848; Remigius, *In Martianum Capellam*, 446.19–21; ps.-Bede, *De mundi constitutione*,
    p. 69, nn. 317–20, 320, 321–7.

sign as it went from Sagittarius to Gemini, then 16 hours less to pass through each successive sign from Gemini to Sagittarius. He proposed several qualitative physical explanations as to why the Sun moves more slowly in its apsis in Gemini, but he clearly lacked the geometric conceptions on which Ptolemaic calculations were based. Yet his simple analysis grows logically from the framework of computus and conforms reasonably well with Martianus's description of astronomical phenomena.

During the ninth century the four astronomical traditions we have been tracing interacted in new ways, subtly transforming Latin astronomy. Alcuin of York and Louis the Pious's advocacy of the feast of All Saints had reflected their apprehension with surviving ritual aspects of the traditional solar calendar, but Helperic of Auxerre used the kind of horizon observations on which that calendar was based to teach computus. Ecclesiastical computus, meanwhile, had become one of the essential components of the education of the clergy embraced by the Carolingian renewal of education. The long-standing tradition of monastic timekeeping continued to be mentioned in monastic legislation emerging from the Carolingian reforms and could only benefit from the teaching of computus and the diffusion of texts on the classical constellations. Those elements of ancient astronomy from Pliny and Isidore that had, since Bede, formed part of the instruction in computus were now joined by Martianus Capella's *De nuptiis* and other texts that taught astronomy as one of the liberal arts. By the end of the century, astronomy had established two secure niches within the curriculum. Side by side with the well-established practical tradition of computus, astronomy had regained its place within the liberal arts, a place from which it would inform the emerging philosophical study of nature.

Yet Latin astronomy still had distinct limitations; it had not achieved the predictive power to which writers on the liberal arts had so confidently alluded. Computus taught the dates of the solstices and equinoxes and could predict the date of the Paschal Full Moon. It also provided rules for computing the position of the Sun and Moon in the zodiac in terms of their average or mean motion. But it could not account sufficiently, as Ptolemy had, for the effect of variations from that mean motion on the Sun's or Moon's position or on the exact date of full Moon. Much the same can be said of the discussions of the periodic motions of the other five planets. These periods were only approximate, given mostly in integral numbers of years, and although retrograde motions were sometimes mentioned there was no attempt to quantify them, let alone the planets' minor non-uniformities. Similarly, while the changing length of day and the changing visibility of the stars were explained in qualitative terms by the spherical shape of the earth and the motion of the Sun through the stars along the inclined path of the ecliptic, this explanation did not lead to any geometrical demonstration or trigonometrical calculation of the length of daylight or the rising and setting of stars. Finally, while medieval astronomical codices now described the general configuration of the stars in the constellations of antiquity, these "star catalogs" gave no celestial coordinates as had Ptolemy and Hipparchus.

One is left with the impression that in the ninth century astronomy did not aim at exact prediction. On the one hand, it served the practical needs of ecclesiastical timekeeping, both in keeping the liturgical calendar and in reckoning the time of night and day for the divine offices. Second, astronomical concepts and symbols were an essential part of a general liberal education and thereby provided a body of knowledge, more literary and artistic than scientific or philosophical, that was shared by the clerical and lay nobility. Yet it would be some time before the Latins would master the predictive astronomy of antiquity, and before this happened they would encounter a new offshoot of that knowledge that had been nurtured in the fertile oases of the Islamic world.

# CHAPTER NINE

# The encounter of Arabic and Latin astronomies

[The astrolabe can be used] to find the true time of day, whether in summer or wintertime, with no ambiguous uncertainty in the reckoning. Yet this seems most suitable for celebrating the daily office of prayer and to be excessive knowledge for general use. How pleasing and seemly the whole proceeds, when with the greatest reverence at the proper hour under the rule of a just judge, who will not wish the slightest shadow of error, they harmoniously complete the service of the Lord.

School of Gerbert (Pope Sylvester II), *On the Uses of the Astrolabe*[1]

By the tenth century the Latin West had two well-established traditions of practical astronomy. Simple observations of the Sun and stars served monastic communities to determine the times for prayer by day and night. Ecclesiastical computus called upon rudimentary arithmetical calculations using approximate lunar and solar periods to regulate the place of Easter, and its related luni-solar feasts, within the Julian solar calendar. While computus had become incorporated into the curriculum of monastic schools, these practical astronomies were only slightly connected with the partly understood survivals of ancient astronomical theory.

Unlike computus, the qualitative geometrical astronomy taught among the liberal arts provided only the most general guide for astronomical observation and even less basis for calculation. In contrast, since the eighth century, Islamic scholars had come to grips with Greek and Indian geometrical astronomy, mastering its theoretical and empirical foundations and incorporating both this theory and their own observations into a wide range of astronomical texts.[2] Most of these texts were far beyond the level of Latin computists and observers. But Islamic scholars had also used these theories and observational data to produce astrolabes and other instruments to determine the time of day and night by observing the Sun and

---

1. *De utilitatibus astrolabii*, 5.4. The text has been variously attributed to Gerbert himself, to one of his students, or to Hermann of Reichenau (1013–1054); see below note 32.

    The metaphor of the astrolabe as a just judge alludes to James 1:17: "All good gifts and all perfect gifts come from above, descending from the Father of light, with whom there is no change nor shadow of error."

2. Kennedy, *Islamic Astronomical Tables*.

stars, as well as to prepare instructions for their use and construction, and to provide simple texts on the calendar and more advanced astronomical tables that would, in time, meet the maturing interests and abilities of Latin scholars.[3]

Not only did the practical focus of these texts and instruments meet the needs and concerns of the Latin West, it also provided a transitional stage between the simple practical astronomies of the Latin West and the more sophisticated theoretical astronomies known in the Arabic world. These practical texts and instruments incorporated elements of astronomical theory and observational data in forms that were both useful and comprehensible to Latin scholars; as such they enriched Latin mastery of those poorly understood elements of ancient astronomical theory. It is to the assimilation of these texts and instruments that we must turn to sketch out the sources of a true renaissance in Latin astronomy.

## Practical astronomies at Córdoba and Gorze

The earliest hint of Latin Christendom's rediscovery of ancient astronomy through contact with Islamic scholars appears in a brief diplomatic contact between the Caliphate of Córdoba in Spain and the monastery of Gorze. Córdoba, lying on the western fringes of the Islamic world, was a vital center of culture. Beginning with 'Abd al-Rahmān III, the Umayyad caliphs had modeled their capital on the Baghdad of the 'Abbāsids. Arabic learning, including the study of both theoretical and practical astronomy, flourished under the Umayyads.

A significant example of Spanish practical astronomy, the Calendar of Córdoba, had been written by two officials of the caliphate: Bishop Recemund of Elvira (known in Arabic as Rabī' Ibn Zayd) and the caliph's secretary, 'Arib Ibn Sa'd al-Katīb al-Qurtubī. The calendar, or Kitāb al anwā' (Book of the anwā'), belongs to a genre of Arabic practical astronomical texts in the tradition of earlier Mediterranean calendars. It provides the kind of rudimentary astronomical data found in Virgil, Aratus, and Pliny, including the meteorological significance and date in the Julian calendar of the heliacal risings and cosmical settings, the anwā', of important constellations; the dates that the Sun enters each sign of the zodiac; the length of daylight; and the height of the Sun and the length of shadows at noon. Bishop Recemund contributed a distinctly Christian element to the book; listed among the agricultural and meteorological lore in the calendar are Christian feasts, much as they appear in the liturgical calendars found throughout Christendom.[4]

This excerpt, spanning the fifteenth to the twenty-fifth of March, illustrates the calendar's range of concerns:[5]

---

3.  Islamic practical astronomies are surveyed in D. King, "Folk Astronomy in Islam."
4.  Calendar of Córdoba, ed. Pellat; Calendar of Córdoba, ed. Martínez Gázquez and Samsó; Kunitzsch, Uber eine anwā'-Tradition, pp. 7–8.
5.  Calendar of Córdoba, ed. Pellat, pp. 56–59; ed. Martínez Gázquez and Samsó, pp. 36–37. My translation follows Pellat's edition of the earlier Latin text, supplemented by his translation from the Latin and Arabic and by the thirteenth-century Latin version.

15 The birth of horses begins now in the maritime regions and continues until the middle of April.

16 According to the experimenter [al-Battānī] the Sun moves from the sign of Pisces into the sign of Aries. The day and night are now of equal length; it is the vernal equinox.

17 The end of wintertime and the beginning of springtime according to the opinion of the computists (*calculatorum*) and astronomers (*equatorum*), and the opinion of Hippocrates, Galen and the learned physicians.

18 The altitude of the Sun at noon is 52.5°, and everything casts a shadow equal to ¾ of its height.

19 Evening twilight lasts 1 2/5 hours after sunset; and morning twilight begins when the same time remains before dawn. This is the average duration of twilight.

20 According to the opinion of the *sindhind*,⁶ the Sun now descends into Aries and according to them it is the equinox.

21 The fourth *magnetis*, and among the Christians {the feast of the incarnation of the Word in the womb of Mary}.⁷

22 Now among the Christians is the feast of the turning of the solar year and it is the beginning of time among them, and the first hour of their Paschal feast, which may not precede that date.

23 [no entry].

24 The rising of the lower spout (*al-Farġ al-Muʾaḫḫar*) in the morning twilight, and it looks like this: The setting of *al-ʿAwwāʾ* [the kennel] at the hour of morning twilight, and it looks like this:⁸

   Now is the beginning of the *nawʾ* of *al-ʿAwwāʾ*, which lasts for three nights. This constellation is similar to the dogs which follow the Lion; and they say that it is the two haunches of the Lion. It is a *nawʾ* of little significance. And opposite it rises the lower spout. This *nawʾ* is the first of springtime and its rain is called *sayyif*.⁹

25 The winds which blow now damage, by their violence, the early figs and the formation of fruits.

   {The Annunciation to the Lady. On this day the Sun was created, the Sun of Justice was conceived, died, and rose from the dead.}¹⁰

6. The *sindhind* refers to the Hindu version of Greek epicyclic astronomy, which stems from a late eighth-century Arabic translation of one of the Sanskrit siddhantas. It and the astronomical tables of Muhammad Ibn Mūsā al-Khwārizmī which employed it were especially influential in Spain. Kennedy, *Islamic Astronomical Tables*, pp. 128–129.

7. The fourth *magnetis* is the fourth of a series of seven weeks bearing this name.
   The bracketed discussion of the Incarnation only appears in the Arabic text; it is omitted in the early Latin version. In the thirteenth-century translation a correct entry for the feast of the Annunciation appears at the traditional Roman equinox, 25 March.

8. This paragraph, with figures illustrating these Arabic constellations, is found only in the early Latin version. Calendar of Córdoba, ed. Pellat, p. 20.

9. *Nawʾ*, the singular of *anwāʾ*, denotes a period of heliacal rising or cosmical setting.

10. The bracketed paragraph occurs only in the thirteenth-century translation.

A few points should be noted about the calendar's astronomy. The agricultural and meteorological advice has a Mediterranean focus; it would not be useful in France or Germany. Although Latins were interested in the risings and settings of constellations, comprehension and use of the *anwā*' presume knowledge of both classical and Arabic constellation names. These sections are not likely to have attracted the attention of readers beyond the Iberian peninsula. But running through this month, as through all the months, are practical data that reflect the influence of geometrical astronomy.

The duration of daylight and nighttime and the height of the Sun at noon and the length of noontime shadows do not follow the linear approximations from classical antiquity found in Gregory of Tours and later computistical collections. They have been computed geometrically for the latitude of Córdoba; their results match those computed using the astronomical tables of al-Khwārizmī, which circulated widely in Spain. Of course, Latins could always dismiss any discrepancy with their traditional values as due to the well-known changes of these phenomena from southern to northern climes.[11]

Less easily dismissed were the various dates given for the vernal equinox (16, 17, 20, and 22 March), for the solstices, and for the entry of the Sun into the twelve signs of the zodiac, which incorporate new data from the *sindhind*, from the Greek medical tradition, and from al-Battānī's observations of 882. All these values differ from the accepted dates of 21 March (according to the Greeks and Egyptians) and 25 March (according to the Romans), which had persisted in Latin ecclesiastical calendars since the time of Bede.[12] The Calendar of Córdoba also tried to follow this tradition, providing a "Christian" equinox, but it confused this equinox with the earliest date of Easter, 22 March, the day following the "Greek" equinox of 21 March. This profusion of dates reflects the increasing divergence between the Julian calendar and the tropical year, and the most recent data reflect the highest levels of Islamic observational technique.[13] Al-Battānī's data, as well as the older and less accurate data of the *sindhind*, all disagreed with the solstice and equinox dates commonly found in Latin sources; these dates would introduce discordant astronomical data into the Latin study of ecclesiastical computus.

The discordance between the traditionally accepted values and the data based on Arabic observations and theories becomes especially apparent in the familiar

11.   Samsó, "Materiales astronómicos en el 'Calendario de Córdoba.' "

12.   Vernet, "Les traductions scientifiques"; Samsó, "Calendarios agrícolas Hispanoarabes y Norte-africanos." The traditional Greek and Roman values are found in two ninth- and tenth-century continental calendars as well as in a wide range of tenth- and eleventh-century English calendars. McCulloh, "*Martyrologium Excarpsatum*"; *Das Sakramentar Wolfgangs; English Kalendars before A.D. 1100.*

13.   Al-Battānī's observations of the Sun not only had identified the extent to which the dates of the solstices and equinoxes had changed since classical antiquity, they were also sufficiently precise to identify changes in the length of each of the four seasons. Willy Hartner, "Al-Battānī," in *Dictionary of Scientific Biography*, vol. I, pp. 508–510.

context of a practical astronomical calendar. Discrepancies of this kind, arising in those practical contexts that had provided the core of Latin astronomical study, would undermine confidence in accepted calendrical data. Through computus, Latins had mastered the problems and techniques of defining a calendar, while their astronomies display a recurring and multifaceted interest in the solstices and equinoxes. The variety of conflicting dates given by both ancient and medieval authorities for these significant events suggested a new locus for later astronomical inquiry, inquiries in which the observational and theoretical acumen of the Islamic world had much to offer.

Personal contacts linking one of the authors of the calendar, Bishop Recemund of Elvira, with a learned monk, John of Gorze, suggest one possible means by which knowledge of Islamic astronomy may have first reached the West. In 953 John was chosen by Emperor Otto I to travel to Spain and treat with Caliph ʿAbd al-Rahmān III to end Islamic incursions into the south of France. Negotiating on behalf of the caliph were Bishop Recemund and Hasdei Ibn Shaprūt, a physician, scholar, court official, and leader of the Jewish community in Córdoba, whose interests included the Jewish luni-solar calendar.[14] John stayed in Córdoba nearly three years, while at a crucial point in the discussions Recemund traveled to the Ottonian court to obtain new instructions for John from the emperor. Recemund stayed much of the winter of 955/956 at the monastery of Gorze.[15]

The monastery of Gorze, originally founded in 748, was reestablished in 933 to become one of the major centers of the tenth-century monastic reform. This new foundation also exemplified the roles of astronomy in a tenth-century monastery. The founding members of this community had been drawn towards the monastic life by their activity in an informal group of clerics and nuns that had met in Metz to study both religious and secular topics.[16]

With the reconstruction of the monastery, including a separate convent for nuns, the community at Gorze had a stable basis to continue its spiritual and scholarly activities.[17] Among this scholarly group was John of Gorze, who had learned computus at Toul, and the deacon Bernacer, whom John's biographer

14.  Hasdei had asked the astronomer Dunas ben Tamin about the calendar, who sent him in reply a treatise on the phases of the Moon. S. Stern, "A Treatise on the Armillary Sphere by Dunas ben Tamin," pp. 373–377 in *Homenaje a Millas Vallicrosa*, vol. 2, 1956, as cited in Destombes, "Un astrolabe carolingien," p. 15, n. 27; Ashtor, *Jews of Moslem Spain*, vol. 1, pp. 159–217.

   Vernet suggests that the young al-Majrītī, who later edited al-Khwārizmī's astronomical tables, was a member of Hasdei's research group. This would tie Hasdei even closer to astronomical circles in Córdoba. Juan Vernet, "al-Majrītī," in *Dictionary of Scientific Biography*, vol. 9, p. 39.

15.  *Vita Iohannis Gorziensis*, 121, 127–130. The negotiations are discussed from Arabic and Latin sources in Ashtor, *Jews of Moslem Spain*, vol. 1, pp. 169–176.

16.  Prinz, *Frühes Mönchtum im Frankenreich*, p. 218; *Vita Iohannis Gorziensis*, 33, 43. The significance of this scholarly community was first noted by Thompson, "Introduction of Arabic Science." More skeptical are Millás Vallicrosa, *Assaig*, p. 208, n. 2, and Borst, *Astrolab und Klosterreform*, pp. 55–57, n. 81.

17.  *Vita Iohannis Gorziensis*, 41.

describes as most studious in the art of computus. There can be little doubt that computus formed a significant part of the studies at Gorze.[18]

John's use of astronomy extended beyond the study of computus. He frequently fasted and stayed up through the night on long vigils; he is described staying up, now kneeling before the altar, now standing, his continual recitation of the psalms sounding like the quiet murmuring of bees. And he would go out under the sky to find the time of night, then return to help prepare or light the lamps.[19] John and his colleagues applied the full range of practical monastic astronomies.

Given the similar interests and astronomical approaches at Gorze and Córdoba, the travels of John of Gorze and Recemund of Elvira probably provided one early channel for the transmission of elements of Arabic astronomy to the Latin West. Yet we have no records of what they learned from each other; there are only shared interests in practical astronomy. Although we could not expect John to understand Islamic theoretical astronomy, at the very least his conversations with the bishop would have led John to recognize its extent. There is no evidence that the contact between John and Recemund brought any astronomical texts or instruments from Spain to Gorze. The Calendar of Córdoba was not completed until after Recemund's negotiations and would not be translated into Latin until the twelfth century.[20]

The eleventh-century catalog of the library at Gorze lists no astronomical treatises based on Arabic learning, although a few astronomical texts from the late Roman tradition are cataloged among books on the liberal arts. There is an excerpt from Cassiodorus on the four liberal arts, which would most likely include the astronomical section from his discussion of the quadrivium,[21] as well as works by Boethius, Macrobius, and Martianus Capella. We also find a collection that is more mythological than astronomical, combining an unnamed treatise on the names of the stars (perhaps one of the Carolingian Aratus texts) with the Sybilline oracles. The astronomy that our librarian places among the mathematical disciplines of the quadrivium reflects the descriptive tradition of the liberal arts rather than the advanced geometrical astronomy that was only beginning to arrive from the Islamic world as he prepared his catalog.

Yet there is another substantial body of astronomical texts dealing with com-

---

18.   *Vita Iohannis Gorziensis*, 18, 24.

19.   *Vita Iohannis Gorziensis*, 80.

20.   The calendar stems from the reign of al-Mustansir (961–976); Vernet, "Les traductions scientifiques," pp. 50–51. A calendar bearing the title of *Liber anoe*, i.e., *Kitāb al-anwā'*, is found in the list of Gerard's translations, compiled shortly after his death in 1187; Richard Lemay, "Gerard of Cremona," in *Dictionary of Scientific Biography*, vol. 16, pp. 173–192.

21.   "Catalogue des manuscrits de Gorze." Welborn first noted the absence of Arabic works at Gorze; Welborn, "Lotharingia," p. 192, n. 19.

There are several forms of partial manuscripts of Cassiodorus's *Institutiones*, yet no manuscript limited to the four quadrivial liberal arts is identified in R. A. B. Mynors, ed., *Cassiodori Senatoris Institutiones* (Oxford: Clarendon Press, 1937).

putus that the librarian cataloged, following the pattern we have seen in Isidore's *Etymologies*, among sacred rather than among secular works. The section on Bede's works mixes his commentaries on the Scriptures and the church fathers with three copies of his *"computus major,"* one copy including figures of the signs of the zodiac. Scattered among other scriptural commentaries and collections of sermons are various computistical texts. None of the library's texts on computus, the most creative area of early medieval astronomy, are found among the books on the seven liberal arts. The librarian at Gorze clearly distinguished the astronomy of computus from the astronomy of the liberal arts. The former was a practical religious discipline, related to liturgy, not unlike the practical astronomies that had been practiced and studied by John of Gorze and expounded in the Calendar of Córdoba.

### The astrolabe

The persons involved in the contact between Gorze and Córdoba are well known, while the outcome of their encounter is uncertain; the reverse is the case with the astrolabe. There are many signs of the influence of the astrolabe's arrival in the Latin West, but historians continue to debate exactly who was responsible for its transmission and when and where it first appeared.[22] Despite the uncertain sequence of events, by 1030 at the latest European scholars possessed astrolabes and were teaching their use, drawing on those texts describing the uses and construction of the astrolabe that had made their way from Muslim Spain to the Latin West.[23]

The astrolabe is commonly thought of as an astronomical instrument, used to observe the position of the heavenly bodies, but it is much more than that. Astronomical instruments are used principally in scientific research or in teaching; the astrolabe was used more like a watch or a sundial. It was a practical timekeeper that combined a simple device for astronomical observations with scales for astronomical calculations. Yet even as a practical observing instrument it reflects a fundamental change. Heretofore most astronomical observations had been qualitative: observing the appearance or disappearance of a star, noting the changing phases of the Moon, marking the general position of one of the planets in relation to the constellations, or tracing the annual motion of the Sun along the local horizon. On the back of the astrolabe (Fig. 25) is a sighting vane with which an observer can measure the angular height of the celestial body using a graduated scale on the circumference of the instrument. This measurement is quantitative,

22. Poulle, "Les instruments astronomiques"; Destombes, "Un astrolabe carolingien."

23. Recent discussions of the history of the astrolabe are found in Borst, *Astrolab und Klosterreform*; Turner, *Astrolabes*; O. Pedersen, *"Corpus Astronomicum,"* pp. 67–69.

    Pedersen saw the arrival of the astrolabe as a crucial event: it provided a new link between observation and theory, which previously did not exist in the Latin West, and posed a challenge to master the new instrument's theoretical basis.

Figure 25. The back of an astrolabe. Note particularly the circle calibrated in signs of the zodiac and the eccentric calendar circle, which can be used to compute the position of the Sun in the zodiac for any given day. CCA no. 186, Smithsonian Institution, negative 77–13841. Used by permission.

Figure 26. The front of an astrolabe. Note particularly the ornamented rete bearing pointers marking the positions of those stars that could be observed to calculate the time of night using the circles of constant elevation (almucantars) on the underlying plate. CCA no. 186, Smithsonian Institution, negative 77–13840. Used by permission.

yielding a measured angle of a given number of degrees, an angle that fits within the geometrical framework of spherical astronomy.

Secondly, the astrolabe is itself a physical representation of that geometrical model, for on the front of the astrolabe (Fig. 26) is a network, or rete, that forms a map of the celestial sphere, a map on which the stars, the apparent path of the Sun, and the solstices and the equinoxes are projected onto a plane. In addition, astrolabes are fitted with one or more interchangeable plates engraved with projections of the local horizon, including altitude and azimuth circles computed for various latitudes where the astrolabe would be used. The observer can compute the time of day or night by rotating the projection of the celestial sphere over the appropriate projection of the terrestrial sphere until the symbol of the observed star or the position of the Sun falls on the circle corresponding to its observed altitude.[24]

Note that as a time-reckoning instrument the astrolabe fits easily within the tradition of practical timekeeping reflected in monastic rules and practice. But the astrolabe gives a new level of precision to nocturnal timekeeping, a precision seldom approached even in daytime by the sundial.[25] Furthermore, its geometrical model allows the ready computation of both the customary unequal hours, which divide the two periods from sunrise to sunset and from sunset to sunrise into twelve hours each, and the equal mathematical hours, which divide the daily revolution of the celestial sphere into twenty-four hours.

For nocturnal prayer the astrolabe offers another advantage. A monastic observer can use the model of the celestial sphere to calculate both when the Sun will rise and also when the Sun will reach the point where twilight begins. Thus by observing the height of any convenient star, the observer could tell exactly when to wake the community for morning prayer.

The timekeeping function is stressed in European astrolabes, which, in addition to the usual scale of degrees to measure the rotation of the heavens, have a scale of hours and fractions of an hour that will directly give the time of day. These equal hour scales were calibrated either in degrees or in increments of three degrees, corresponding to a traditional computistical unit, the lunar point, which equals one-fifth of an hour. Besides the principal scales on the front, several auxiliary scales are found on the back (see Fig. 25) of Arabic astrolabes from Spain and on European astrolabes.[26] The first of the two scales that concern us, the

---

24.  For further details of the procedure, see, for example, North, "The Astrolabe"; Turner, *Astrolabes*, pp. 7–9.

25.  A lack of precision in placing the hour lines on many existing sundials is indicated by the inconsistent values of intended latitude derived from measurements of different hour lines on the same dial. Gibbs, *Greek and Roman Sundials*, pp. 16–19, 33, 75, 79.

26.  The division into points antedates the introduction of the astrolabe; in his discussion of the Moon's motion through the zodiac, Bede noted that "each sign has ten points, i.e., two hours, . . . hence five points make an hour." *De temporum ratione*, 17.3–5. Taking five points in an hour, each point thereby represented twelve minutes of time or three degrees of arc.

   For a detailed discussion of the scales typical of European instruments and of Arabic instruments from the Maghrib (i.e., Spain and North Africa) and the Mashriq (the Orient), see Gibbs and Saliba, *Planispheric Astrolabes*, pp. 2–5, 25, 30–47, 52–54.

calendar scale, uses an eccentric circle to trace the annual motion of the Sun through the zodiac according to the Julian calendar. With it the observer could compute the Sun's position for any date without auxiliary astronomical tables. This scale is not found on oriental astrolabes, where the purely lunar Islamic calendar predominated and the calendar scale would serve no useful purpose.

The second scale is composed of a set of curves to measure the unequal hours of daylight directly from an observation of the height of the Sun. The procedure using these curves is simpler, if significantly less accurate, than that using the front of the instrument; this duplication reflects the importance of the measurement of time among the uses of the astrolabe.[27]

One of the first written indications of the astrolabe in the Latin West comes from the Spanish March, a Christian enclave on the borders of Islamic Spain. Late in the tenth century an unknown writer, possibly Archdeacon Lupitus of Barcelona (fl. ca. 975–995), wrote a preface to a lost treatise on the astrolabe.[28] The preface praises the nobility and usefulness of astronomical study, noting the physical dominance and moderating influence of the aetherial fire over the terrestrial elements, the influence of the Moon on the tides, and the significance of the star at Christ's birth. Nonetheless, the author rejected astrologers' attempts to forecast a person's fate from the state of the stars at the time of conception or birth as a frivolous superstition, since a person's fate is wholly committed to divine disposition. The nobility of astronomy was presented in Platonic terms as drawing the mind to the perception of the invisible through contemplation of the visible spheres; its uses are given as the determination of the date of Easter and of the time to chant the divine office by night and day. The author then noted that the text that follows, drawn from Arabic sources, discusses both the uses of the astrolabe and the method of its construction.[29]

The studies of Gerbert of Aurillac (ca. 945–1003), later Pope Sylvester II (999–1003), provide another link with Hispano-Arabic science and the introduction of the astrolabe as the principal new timekeeping device. From 967 to 970 Gerbert studied mathematics in the Spanish March under Bishop Atto of Vich. During much of his active life as abbot of Bobbio and bishop of Rheims he passed this learning on to a succession of students. Gerbert's teaching of mathematics was not

27. Noting that only one medieval astrolabe in a sample of 41 having unequal hour curves had graduations on its alidade to be used in conjunction with them, North questions whether these curves were actually used. I find his scepticism overdrawn, since these curves could have been used without such graduations by marking the noontime height of the Sun. North, "Astrolabes and Hour-line Ritual."

28. Lupitus is mentioned in the correspondence of Gerbert of Aurillac. In 984 Gerbert requested from Lupitus a copy of a translated treatise on *astrologia*. Gerbert, *Epistola 24* in Gerbert, *Opera mathematica*, pp. 101–102; translated as Letter 32 in Gerbert, *Letters*, pp. 69– 70.

    On Lupitus, see Harriet Pratt Lattin, "Lupitus Barchinonensis," *Speculum*, 7(1932):58–64; on the prologue, see Millás Vallicrosa, *Assaig*, pp. 187–189.

29. Lupitus of Barcelona(?), *Fragmentum de astrolabio*.

merely an abstraction, but clearly extended into astronomical measurements, which he saw as the key for comparing "theoretical and actual measurements of the sky."[30]

Various instruments and techniques for timekeeping provide one continuing strand of Gerbert's astronomical concern. He described how to compute tables (*horologia*) for the varying length of day and night through the year using Martianus Capella's arithmetical method; however, the two examples he included reveal all too clearly that not even Gerbert correctly understood Martianus's procedure. He did not know any geometrical method to compute the ratio of day to night at the solstices, yet he gave practical advice for determining this ratio with a water clock to provide the basis for computing a *horologium* for any specific place.

He also constructed a *horologium arte mechanica compositum* during his extended stay at Rheims, another *horologium* for the monastery at Magdeburg used to determine the time by observing the turning of the stars around the pole, and a water clock at Ravenna where he was archbishop from 998–999. These accounts reflect the interest of Gerbert and his contemporaries in using mechanical devices for timekeeping, an interest that would reach new levels of quantitative precision with the arrival of the astrolabe.[31]

The author of the *De utilitatibus astrolabii* (variously attributed to Gerbert, to one of his students at Rheims or Liège, or, least likely, to Hermann of Reichenau) notes that astrolabes were especially appropriate for monastic timekeeping. He praised the astrolabe as a "good gift from above," one that was especially suited to regulate the celebration of the divine office.[32]

The astrolabe soon made its way into the curriculum of monastic and cathedral schools. A recently discovered manuscript fragment from the vicinity of the monastery of Reichenau includes several chapters, one of them numbered the forty-eighth, of an extended treatise on the astrolabe. The surviving chapters are excerpts from earlier astrolabe treatises and concern the stars of the astrolabe and

---

30.  Gerbert, *Letters*, pp. 3–20. *Constantino suo Gerbertus scolasticus*, in Gerbert, *Opera mathematica*, pp. 6–8; cf. Gerbert, *Letters*, Letter 7, pp. 45–47.

31.  Gerbert, *Epistola de horologiis duorum climatum ad fratrem Adam*; Thietmar of Merseburg, *Chronicon*; William of Malmesbury, *De rebus gestis regum Anglorum*, 2.168; *Gesta epp. Halbertstadensium*, C.a. 1209–1218; in Gerbert, *Opera mathematica*, pp. 38–41, 382, 388, 391. Cf. Gerbert, *Letters*, pp. 189–191.

Bubnov and Bergmann interpret both the Magdeburg and Rheims accounts as referring to astrolabes. Wiesenbach argues convincingly that the Magdeburg *horologium* was patterned after Pacificus of Verona's invention. Bergmann, *Innovationen im Quadrivium*, p. 159; Wiesenbach, "Pacificus von Verona."

32.  *De utilitatibus astrolabii*, 1.2, 5.4, 6.1. This treatise is first found in a late tenth- or early eleventh-century collection of astronomical works made for use in the monastery of Santa Maria de Ripoll in the Spanish March. For differing views on the question of attribution, see David C. Lindberg, "The Transmission of Greek and Arabic Learning to the West," in his *Science in the Middle Ages*, pp. 60–61; Bergmann, *Innovationen im Quadrivium*, pp. 148–155; Borst, *Astrolab und Klosterreform*, p. 48; and their sources.

how to observe the Sun or those stars to determine the time of day or night. This text, which has been dated on paleographical grounds to around the year 1000, reflects the incorporation of the astrolabe into the study of astronomy at Reichenau.[33]

Knowledge of the astrolabe spread widely during the first part of the century. One of Gerbert's students, Fulbert of Chartres (ca. 960–1028), the noted master of the school and later (1006–28) bishop of that city, composed a brief mnemonic rhyme placing eight of the astrolabe stars (with their Arabic names) within the familiar zodiacal constellations. A list of the same eight "stars of the hours" appears in the *De horologio secundum alkoram*, a treatise on the astrolabe that had also been quoted by the anonymous Reichenau compiler. Fulbert clearly selected these stars for actual observation; all but two are first magnitude stars and they are well distributed around the zodiac so several will be above the horizon at any time.

> Aldeberan stands out in Taurus, Menke and Rigel in Gemini,
> and Frons and bright Cabalazet in Leo.
> Scorpio, you have Galbalagrab; and you Capricorn, Deneb.
> You, Batanalhaut, are alone enough for Pisces.[34]

Fulbert also prepared a brief glossary giving the equivalent Arabic and Latin names of the parts of the astrolabe. But even as practical astronomy these samples of Fulbert's teaching at Chartres are not very advanced; they give no sign of either Arabic astronomical theory or Latin computus.[35]

A collection of letters exchanged around 1025 between two of Fulbert's former students, Ragimbold of Cologne and Radolf of Liège, reveals serious limits to their education. There is a striking absence of the basic geometrical knowledge required to master astronomical theory. The two scholars, who by that time were masters of schools in their respective cities, had been puzzled by a reference in Boethius to the interior angles of a triangle. Much of their correspondence was a debate over what this term could possibly mean. Yet Radolf wrote an excited note to Ragimbold that he had acquired an astrolabe. He bubbled over with enthusiasm, declaring that Ra-

---

33. Borst, *Astrolab und Klosterreform*, pp. 44–47, 114–117.
34. Frederick Behrends, ed., *The Letters and Poems of Fulbert of Chartres* (Oxford: Clarendon Press, 1976), pp. xvi, xxvii–xxviii, 260–261; Frederick Behrends and Michael McVaugh, "Fulbert of Chartres' Notes on Arabic Astronomy," *Manuscripta*, 15(1971):172–177.

     Aldeberan is α Tauri; Menke (*mankib al-gawza*, the shoulder of Orion), α Orionis; Rigel, β Orionis; *Frons leonis*, α Geminorum; Cabalazet, α Leonis; Galbalagrab, α Scorpionis; Deneb, δ or τ Capricornis; and Batanalhaut (*galb al-hut*, the heart of the fish), β Andromedae.

     Millás Vallicrosa, *Assaig*, pp. 185–187, 292; Kunitzsch, *Typen von Sternverzeichnissen*, p. 4.
35. The eleventh-century manuscript catalog of the monastery of St. Peter at Chartres includes no texts on computus. The only scientific works are rudimentary: Bede's encyclopedic *De natura rerum* and Boethius's *Arithmetica*. *Catalogi bibliothecarum antiqui*, pp. 144–145.

gimbold must see one to appreciate it, but he was unwilling to part with his as it was the model from which he intended to have a copy made. He urged Ragimbold to meet him on St. Lambert's Day to examine it.[36]

Given Radolf's limited mathematical knowledge, he certainly could not have understood the geometrical theory of the astrolabe. Treatises on the theory of stereographic projection would not appear in the Latin West for a century, with the translation of Ptolemy's *Planisphaerium* in 1143.[37] Nevertheless, Radolf apparently knew enough about the use of the astrolabe to want one and enough about the practical details of its construction to attempt to have one made.

Even before the first theoretical analyses of the astrolabe appeared in Latin, instructions for constructing the instrument circulated. The practical geometry and mechanical skill that these treatises demanded of medieval metal workers, which they readily transferred from creating artistic ornaments for church and court to making precise instruments for observation and computation, are not our immediate concern. More significant are those aspects of astronomical knowledge included in texts describing the astrolabe.[38]

Details of mathematical astronomy are implicit in the simple instructions for the geometrical projection of the celestial sphere, a project which at first glance scarcely goes beyond the kind of artistic models that we have seen in Carolingian manuscript illustrations. Yet the astrolabe is not merely an attractive ornament providing a qualitative picture of the heavens; since it was used for astronomical observations and computations, it must be a mathematically precise instrument, incorporating quantitative data about the heavens.

Construction of an astrolabe necessarily defined the tropics, the celestial equator, and the zodiac in precise quantitative terms, in contrast to those texts of the antique tradition which described them qualitatively. Thus three concentric circles were divided into four equal quadrants by two lines (the colures) passing through the solstices and equinoxes, and each quadrant was divided into 90° making a whole circle of 360°. The signs of the zodiac were fixed as arcs of exactly 30°, with the equinoxes and solstices unambiguously defined according to the astronomical norm at the beginning of Aries and Libra, Cancer and Capricorn. The circles marking the turning points of the Sun were projected 24° north and south of the equator, using a simple geometric construction.[39] Once these circles were

36.   "Une correspondance d'écolâtres," pp. 234–240, 283; MacKinney, *Fulbert and the School of Char-tres*, pp. 14–15, 29–30; Southern, *Making of the Middle Ages*, pp. 198, 201–203. That Ragimbold still did not know what an astrolabe looked like suggests that Fulbert's teachings on astrolabe stars and familiarity with the parts of the astrolabe came late in his career.

37.   Thomson, *Jordanus de Nemore*, pp. 50–60.

38.   *De mensura astrolapsus; De mensura astrolabii*; Hermann of Reichenau, *De mensura astrolabii*; ps.-Bede, *Libellus de astrolabio*; Ascelinus, [*Quomodo fiat astrolabium*].

39.   Hermann of Reichenau, *De mensura astrolabii*, pp. 204–205. Later writers would draw on astronomical texts to suggest the unachievable construction of an angle of 23° 33' 30"; Raymond of Marseille, "Traité d'astrolabe," pp. 878–79.

defined, the inclined circle of the ecliptic was constructed tangent to the two tropics and intersecting the equator at the equinoxes. The equator, the ecliptic, and the signs of the zodiac provided a set of coordinates against which the bright stars could be plotted. And so the construction texts generally list the names and coordinates of the stars to be placed on the rete of the astrolabe.

This whole procedure reveals a striking contrast with the astronomy of the Carolingian era. Where late antique writers had talked in general terms of the circles of the heavens, astrolabe texts gave specific procedures for constructing those circles and precisely dividing them. Where Carolingian "star catalogs" had counted the brighter stars in each part of the constellation, these texts provided precise quantitative data. Where antique authors had been verbose and medieval scribes had decorated their texts with splendid drawings of the constellations, these texts succinctly listed the coordinates of the astrolabe stars and sometimes illustrated their locations on the face of an astrolabe. Although the astrolabe texts commonly gave the coordinates of only 27 stars compared to Ptolemy's 1022, we are closer in quantitative presentation to Ptolemy's star catalog and closer in observational concern to Gregory of Tours's *De cursu stellarum* than we are to the Carolingian texts.[40]

Besides the coordinates and the stars, some construction texts provided another significant piece of astronomical data. Since the back of many astrolabes included a calendar scale to determine the position of the Sun in the zodiac for any day in the year, instructions for this scale incorporated a crude model of the Sun's annual motion. The descriptions in the early construction texts are rudimentary, with no hint of advanced astronomical data. Two of the earliest texts provide an extremely simple scheme, in which each month is arbitrarily taken as thirty days and matched to a sign of thirty degrees, but ordered in such a way that the beginning of each sign falls in the middle of each month. In this system the solstices and equinoxes mark the sixteenth of these schematic thirty-day "months."[41] Hermann of Reichenau's *De mensura astrolabii* employed a more advanced scheme that gave each month its proper length in a circle divided uniformly into 365 equal parts.

Hermann follows certain unnamed *moderni* to place the vernal equinox on 18 March. These moderns do not appear to be Arabs, for the eighteenth does not appear among the various dates for the vernal equinox given in the Calendar of Córdoba. However, this is the date given for the Sun's entry into Aries in almost all Latin martyrologies, dating back to 818.[42] Now that the astrolabe and its texts identified the vernal equinox with the entry of the Sun into Aries, the discrepancy

40. *De mensura astrolabii*, pp. 300–302; Hermann of Reichenau, *De mensura astrolabii*, p. 209; Ascelinus, [*Quomodo fiat astrolabium*], pp. 224–225; Kunitzsch, *Typen von Sternverzeichnissen*; cf. Ptolemy, *Almagest*, 7.5–8.1.

41. *De mensura astrolabii*, pp. 298–299; *De astrolabii compositione*, pp. 312; *De astrolabii quadrato*, pp. 200.

42. McCulloh, "*Martyrologium Excarpsatum*"; Bernold of Constance, "Kalendar"; *Das Sakramentar Wolfgangs*; *English Kalendars before A.D. 1100*.

between the date of the entry into Aries and the traditional dates for the equinoxes according to the Greeks (21 March) and the Romans (25 March), which appeared on the same page of so many liturgical calendars, became glaringly obvious.

One indication of the increasing confusion appears in the liturgical calendar devised by Hermann's successor as chronicler at Reichenau. In 1074 Bernold prepared a calendar that included new dates for the solstices and equinoxes on the sixteenth of the month, again following the *moderni*. These new dates are not Hermann's, but agree with those earlier astrolabe texts in placing the solstices and equinoxes at the middle of each month. Standing in opposition to the solstice and equinox dates of the *moderni* were the traditional Greek and Roman solstices and equinoxes, and dates for the Sun's entry into each sign taken from earlier liturgical calendars.[43]

## All things in number and measure

Although the astrolabe was used in the Latin West since early in the century, it is not until 1092 that we find a record of a quantitative astronomical observation carried out to improve astronomical predictions.[44] About an hour before dawn on 18 October, Walcher, a Lotharingian monk who was prior of the monastery of Great Malvern in England, saw the Moon begin to grow dark. For some time Walcher had been dissatisfied with existing computistical tables, which seemed inadequate as a guide for those human activities, such as medical practice, that depended on the waxing and waning of the Moon.[45] He wanted to find the precise time of a lunar eclipse as the basis for a new set of lunar tables that would give the exact time of each new Moon. Taking his astrolabe, he measured the height of the Moon at its darkest and found it to be fifteen degrees above the western horizon. Since the eclipsed Moon was on the zodiac and directly opposite the Sun, Walcher used this single observation to determine the exact time of the lunar eclipse, between the third and fourth points of the eleventh unequal hour of the night (10:45–11:00).[46] He then converted the times of the eclipse and of

43.  Bernold of Constance, "Kalendar."
44.  Oxford, Bodleian Library, MS Auct. F.1.9, ff. 86r–96r; Southern, *Medieval Humanism*, pp. 166–168; Haskins, *Studies*, pp. 113–117.
45.  MS Auct. F.1.9, fol. 86va.
46.  Contrary to the usual practice of computists, who measured the motion of the Moon using lunar points of a fifth of an hour, Walcher used solar points of a quarter of an hour for his lunar tables and his eclipse observations; MS Auct. F.1.9, ff. 86vb, 87ra, 93v.

    Poulle relied on Haskins's incomplete published text when he maintained "without hesitation" that Walcher only used his astrolabe for computations and determined the time by observing the polar constellations with a nocturnal. Poulle, "Walcher de Malvern," pp. 53–54. In fact, Walcher tells us that he determined the time by measuring the height of the Moon: "Tunc vero nullius alius stelle altitudinem inveniende hore necesse erat inspicere, quam ipsius lune et ipsius altitudo xv$^{cim}$ graduum erat." MS Auct. F.1.9, fol. 90rb.

sunset to equal hours and calculated that the eclipse occurred at the third point of the thirteenth equal hour after sunset (12:45).[47]

From this one eclipse observation, Walcher easily computed the time of the preceding astronomical new Moon, which he required for his new lunar tables, as 3 October, 19:30 hours. Walcher then computed tables giving the exact time of each new Moon (reckoned in hours and points from sunset)[48] over four nineteen-year cycles covering the seventy-six years from 1036 through 1111. Walcher chose a beginning date some sixty years before his observation so that his tables would begin with the year when the new Moon occurred just after sunset on the night of 1 January. Thereafter, he hoped, the tables would be perpetual.

Walcher's investigation drew on two medieval astronomical traditions, time-keeping and computus, but with decidedly new purpose and precision. Heretofore lunar tables were principally concerned with the Paschal Full Moon in order to determine the date of Easter. Consequently, they were adequate if they found that one full Moon to the nearest day. Their concern with Easter Sunday led to their arrangement in a 532-year cycle incorporating the 19-year luni-solar cycle, the 4-year cycle of leap years, and the 7-year cycle of the days of the week. Walcher's tables are not traditional computistical tables or ecclesiastical calendars; Walcher notes that his computations will not give the liturgically correct value of the Paschal Moon.[49] This abandonment of liturgical questions allows Walcher's tables to be simpler. Since they are not concerned with the date of Easter Sunday, they can be made cyclic after only 76 years, a period combining the 19-year luni-solar cycle and the 4-year cycle of leap years.

Walcher's tables present the time and date of each new Moon, although he recognized that the tables were only as good as the eclipse observation that fixed the origin of the lunar cycle. He properly judged that his observations with the astrolabe were accurate only to the nearest point, that is, fifteen minutes.[50] Thus he limited his lunar tables to the nearest point rather than apply the ever smaller fractions found in some computists' abstract discussions of time.

By increasing computistical precision a hundredfold from the nearest day to the nearest point, Walcher posed new astronomical problems. His computations

47. Walcher's astrolabe apparently lacked scales to read equal hours from sunset directly. These are found on Arabic astrolabes from the East but not on Spanish or Latin astrolabes. Gibbs and Saliba, *Planispheric Astrolabes*, pp. 5, 53–54.

48. Walcher's reckoning of equal hours from sunset might seem anomalous, since it measures time from the varying norm of sunset rather than the more uniform norms of midnight or noon. Yet this choice is appropriate for his concern with tables for the new Moon, which appears at sunset, a concern like that of the Islamic world where months are reckoned from the observed new Moon and days from sunset.

49. MS Auct. F.1.9, fol. 91rb.

50. MS Auct. F.1.9, fol. 91rb. Walcher's calculated time from sunset corresponds to 05:42 local apparent time; recomputations from Walcher's observation give a result of 05:22 local apparent time; modern calculations place the time of the opposition at 05:00 GMT (05:07 local apparent time at Malvern). Goldstine, *New and Full Moons*.

considered only the average length of the month, for which he used the relationship established by earlier computists that 235 lunar months occur in 19 years,
or 6939 days, 18 hours. This yields an average value for the mean lunar month
of 29.53085 days, which Walcher expressed in solar points of a quarter of an hour
as 29 days, 12 hours, and 3 points, less 9 points in 19 years.[51] But this average is
not constant; the Moon moves with a nonuniform speed in its orbit, being as
much as five degrees ahead or behind its mean motion at new or full Moon.
Similarly the Sun anticipates or lags behind its mean apparent path through the
zodiac by as much as two degrees.[52] These two anomalies can cause the time of
an eclipse to vary by more than thirteen hours from times computed using the
traditional mean motions of ecclesiastical computus. In the past, when computists
wished only to find the day of the Paschal Moon, they could ignore such minor
observational anomalies; Walcher could not.

Over the years as Walcher compared the results of his tables with eclipses that
he either observed or learned about from others, he assembled a disturbing collection of disagreements. His tables listed a new Moon for the first point of the
fourth hour of the night of 22/23 November 1093; the true conjunction was revealed by a solar eclipse at the third hour of the next day. The tables indicated a
full Moon an hour before sunset on 10 January 1107; Walcher observed an eclipse
at the fifth hour of the night of 10/11 January. Twelve months later the tables
indicated a full Moon for the eighth hour of the night of 31 December 1107/1
January 1108; the Moon rose eclipsed at sunset on 1 January, which disagreed "by
more than sixteen hours from the order of the [lunar] cycle."[53] By comparing these
observations with his tables – tables which incorporated the computists' assumption
of uniformly long months – Walcher rediscovered the inconstancy in the course
of the Moon. Since this discovery demanded a theoretical understanding of these
anomalies, Latin astronomers needed a different kind of astronomy.

Independent discovery proved unnecessary, for such an astronomy already ex-
isted in the Islamic world and would soon take root in a circle of West Country
mathematicians whose most prominent members were Walcher of Malvern, Peter
Alfonsi, and Adelard of Bath.[54] Like Walcher, Peter Alfonsi, a converted Spanish
Jew and physician to Henry I of England, praised the usefulness of the astronomical art for medicine, "which cannot be fully mastered except through astronomy." Astronomy can determine the correct time for cauterization, the lancing
of abcesses, and bleeding. He also shared Walcher's respect for real experience,

51. MS Auct. F.1.9, fol. 87rb. Elsewhere Walcher expressed the length of a month in lunar points
    of a fifth of an hour as 29 days, 12 hours, 3 points, 1 minute, 1 moment, and 348 atoms; fol.
    86ra.
52. I have only considered that portion of the solar and lunar anomalies due to their eliptical orbits;
    I have ignored the influence of the Sun on the Moon's motion, which is not significant at
    conjunction and opposition.
53. MS Auct. F.1.9, ff. 95vb –96ra.
54. For Adelard and his colleagues, see Gibson, "Adelard of Bath"; Dickey, "Adelard of Bath."

dismissing the mere book learning of those who presumed that they could learn astronomy by reading Macrobius and other classical sources.[55]

Since Walcher's lunar tables and eclipse observations reflect the medical concerns and observational approach of "his master" Peter Alfonsi, it is not surprising that after completing his tables we find him presenting Peter's teachings on the lunar nodes. The concept of the lunar nodes, that is, the ascending node where the Moon crosses the ecliptic from south to north and the descending node where it crosses from north to south, refines the common understanding that the Moon is usually either north or south of the ecliptic into part of a quantifiable geometrical model of the heavens. Since medieval astronomers had long known that eclipses can happen only when the Moon is on the ecliptic, the theory of the nodes provided a method for determining the possibility of eclipses in terms of the relative motions of the Sun, Moon, and lunar nodes, which Walcher continued to express in terms of simple mean motions through the zodiac.[56]

Walcher presented Peter's values for the mean daily motion of the Sun, Moon, and lunar nodes, apparently derived from al-Khwārizmī's tables,[57] and showed how to calculate the mean motion in days, weeks, months, and years. But from his own anomalous eclipse observations he knew that mean motions are not adequate, noting that the greater and lesser motions of the Moon differ from the mean by as much as six degrees.[58] He also noted Peter's dissatisfaction with the traditional "book values" for the summer solstice of 24 and 20 June, when actual observations indicated 16 or 17 June, presenting an alternative theory that the solstices and equinoxes were alternately retarded and advanced every nine hundred years.[59]

Walcher is an important transitional figure; he knew, and was dissatisfied with,

55. Petrus Alfonsi, *Epistola*; translated in Hermes, *Petrus Alfonsi*, pp. 69–70.

56. Oxford, Bodleian Library, MS Auct. F.1.9, ff. 96r–98v; Millás Vallicrosa, "Pedro Alfonso," pp. 87–97. Southern, *Medieval Humanism*, p. 170; Haskins, *Studies*, pp. 116–119.

57. Walcher's mean daily motion for the Moon (13;10,24,52) appears to be a misreading of al-Khwārizmī's 13;10,34,52 . . . ; his value for the lunar nodes (0;3,10,50) is close to al-Khwārizmī's 0;3,10,48 . . . ; and his value for the Sun (0;59,8,15) falls in the range of values (0;59,8,10 to 0;59,8,21) that could be inferred from al-Khwārizmī's tables or from many other sources. al-Khwārizmī, *The Astronomical Tables*, pp. 90, 92, 95, 148, 156, 178.

58. Millás Vallicrosa, "Pedro Alfonso," p. 89; MS Auct. F.1.9, fol. 96vb. This discussion of the variation of the motion of the Moon shows the influence of Arabic mathematical astronomy. The $\pm 6°$ variation in the motion of the Moon exceeds the maximum equation ($\pm 4°56'$) in al-Khwārizmī's *zīj* and is less than the maximum combined value of the first and second inequalities ($\pm 7°40'$) found in Ptolemy's *Almagest* or *Handy Tables*. al-Khwārizmī, *Die astronomischen Tafeln*, tab. 23; Ptolemy, *Almagest*, 5.8. At eclipses, i.e., at new and full Moon, the difference between Ptolemy and al-Khwārizmī becomes negligible (less than 0°5'). Neugebauer, *HAMA*, pp. 80, 93–94, 988, 1106–1108.

59. Millás Vallicrosa, "Pedro Alfonso," pp. 92–93; MS Auct. F.1.9, ff. 97rb–97va. A similar oscillatory motion, or "trepidation of the equinoxes," is mentioned in a marginal note to Peter's version of al-Khwārizmī's astronomical tables, but the period mentioned here does not appear in that, or any other common discussion of trepidation. al-Khwārizmī, *The Astronomical Tables*, pp. 182–184; Neugebauer, *HAMA*, pp. 631–634.

the limited precision of existing computistical tables. He used his own observations, made with an astrolabe, as the basis of a new and more precise set of lunar tables. Discrepancies between his tables and his further observations raised questions that he tried to answer in terms of the smatterings of Arabic astronomy he had learned from a Spanish colleague. He did not, however, go beyond these vague discussions to compute the true positions of the Sun and Moon on geometrical principles, mentioning only that Peter's books, which could have provided this information, were then overseas.[60]

The missing books mentioned by Walcher apparently included the astronomical tables (or *zīj*) of Muhammed Ibn Mūsā al-Khwārizmī, which Peter partially translated in 1116. Al-Khwārizmī's *zīj* suited the prevailing level of astronomical knowledge in the Latin West. Like Ptolemy's *Handy Tables, zījes* presented astronomical tables and practical instructions, or *canones*, for their use, with scant information about their theoretical basis. First, and most important, were the tables for the mean motions of the Sun, Moon, and other planets and tables for correcting these mean motions to obtain the bodies' true positions; they were followed by additional tables to find the length of day, compute horoscopes, and perform other astronomical and astrological computations.

The eager reception of this astronomy is illustrated by the many Latin astronomical tables derived, in whole or in part, from al-Khwārizmī's *zīj*. The widely circulated translation by Adelard of Bath tabulated the mean motions in the lunar years of the Islamic calendar, counting from the epoch of the Hijra. The earlier, partial, version by Peter Alfonsi revised the Arabic tables so that the mean motions were recomputed, somewhat carelessly, in Julian years reckoning from 1 October 1116.[61] In the late 1140s Robert of Chester edited Adelard's translation, altering the sequence of text and tables, replacing Arabic technical terms with their Latin equivalents, and making other minor changes, such as adding notes for using the tables at the meridian of London. In the same decade Raymond of Marseilles recomputed the *Tables of Toledo* for the longitude of Marseilles. These were among the first of many collections of astronomical tables computed for different cities that emerged early in the century, drawing an eclectic and not always consistent mix of elements from the Hindu astronomy of al-Khwārizmī, the Ptolemaic astronomy of al-Battānī, and al-Zarqālī's work in the *Tables of Toledo*.[62]

Looked at from the perspective of the early 1100s, al-Khwārizmī's *zīj* transformed astronomical computations, yet in a comparatively moderate way. Walcher, like the earlier computists, had been concerned with the time of eclipses,

60.  Millás Vallicrosa, "Pedro Alfonso," p. 89; MS Auct. F.1.9, fol. 96va.
61.  Al-Khwārizmī, *Die astronomischen Tafeln*; al-Khwārizmī, *The Astronomical Tables*.
62.  Raymond Mercier is currently engaged in a project to establish the relations among some of these tables. Mercier, "Astronomical Tables." See also Dickey, "Adelard of Bath," pp. 35–60; Zinner, "Tafeln von Toledo"; Toomer, "Survey of the Toledan Tables"; Toomer, "Solar Theory of az-Zarqāl," p. 308; "Solar Theory of az-Zarqāl: An epilogue."

which were computed using rudimentary arithmetical techniques. For greater accuracy Walcher needed to find the true positions of the Sun and Moon and, from these, the exact time of conjunction and opposition. Al-Khwārizmī's *zīj* provided tables and simple arithmetical procedures to find these true positions, based on the geometrical models of the *sindhind*, a Hindu version of a simpler, pre-Ptolemaic Greek astronomy. This astronomy's underlying model for the Moon has only one epicycle, and consequently, al-Khwārizmī tabulates only one correction to the Moon's mean motion. This system is easier to use, employs fewer tables, and, other things being equal, would not differ substantially from a more accurate Ptolemaic model at the crucial times of conjunction and opposition. Whether through chance or choice, it was this simpler form of mathematical astronomy that first answered the quest of European scholars for more accurate lunar predictions.

Even more important, with these tables European astronomers could now accurately compute the positions of Mercury, Venus, Mars, Jupiter, and Saturn. As with the Moon, al-Khwārizmī's planetary tables provided a good starting point, for the geometrical models of the *sindhind* led to simpler computational techniques than those of Ptolemy's *Almagest*.[63] These techniques, coupled with tables and instruments such as the astrolabe for computing the risings and settings of the stars and planets, also contributed to the emergence of that other practical astronomical art, horoscopic astrology.

Somewhat surprising, given Peter's interest in celestial influences, is the absence of the astrological tables from his early version of the *zīj*. More consistent with Peter's approach to astronomy is the omission of the trigonometric tables. Like Walcher, Peter still saw predictive astronomy as an arithmetical art akin to computus; he considered the geometrical part of astronomy to be concerned with the nature, number, and extent of the heavenly spheres, whereas their movements were reckoned using simple arithmetic. While he knew the arithmetical techniques for using Arabic astronomical tables, there is no sign that he had mastered their geometrical foundations.[64]

Adelard of Bath surpassed Peter by providing the full text of al-Khwārizmī's *zīj* and, in his later writings, he displays a deeper comprehension of the geometrical principles of astronomy. This should not be surprising; after studies at Tours, around 1112 Adelard went to southern Italy and Sicily where he learned Arabic. He wrote on a broad range of scientific and mathematical topics, as well as translating Abū MaʿShar's *Lesser Introduction* to astrology[65] and producing one of the

63. Neugebauer, "Transmission of Planetary Theories."

64. Al-Khwārizmī, *The Astronomical Tables*, pp. 133–234; Petrus Alfonsi, *Epistola*, p. 105; Hermes, *Petrus Alfonsi*, p. 80.

65. Gibson, "Adelard of Bath," pp. 9–13; Charles Burnett, "Adelard, Ergaphalau and the Science of the Stars," pp. 133–145 in Burnett, *Adelard of Bath*, here pp. 133–135.

earliest full translations of Euclid's *Elements* of geometry from Arabic into Latin.[66] Thus when called upon to write a primer on the astrolabe for young Henry Plantagenet (later Henry II of England) Adelard wrote a quite different treatise from those we have seen before.[67]

The uses of the astrolabe discussed in Adelard's treatise betray the growing influence of astrology and the concomitant theory of celestial influences reflected in Adelard's translations and in the works of Walcher and Peter. Conversely, there is less emphasis on the routine use of the astrolabe to keep time. These emphases may reflect the interests of court circles, which Adelard frequented, as well as the historical rise of astrology.[68] Even more striking is the increasingly technical emphasis on geometrical astronomy. Adelard went beyond a discussion of the celestial spheres, for which the astrolabe provides a physical model, to a consideration of the geometrical principles underlying the computation of planetary positions. He showed how to use an astrolabe to measure the approximate position of a planet in the zodiac, based on a simple observation of the altitude of the planet and a comparison star. For more exact positions of the planets he directed his reader to the *zījes* of the Arabs, criticizing the naive interpretations of those (here we may see Walcher and the computists) who assumed that the planets travel through each sign of the zodiac in an equal amount of time. He explained the retrograde motion of the five planets in terms of epicycles but followed al-Khwārizmī's model in which the Moon and Sun, which do not undergo retrograde motion, lack epicycles. Adelard did not write a mere users' guide to the astrolabe; he gave his future king an introduction to the new astronomy.[69]

Throughout the twelfth century the interest in the new astronomies found in Arabic astronomical tables and instruments both reflected and contributed to a widespread, if ill-defined, dissatisfaction with Latin astronomical traditions. Observations like Walcher's, applying a new astronomical instrument to bring increased observational precision to familiar astronomical questions, provided one challenge to the inherited traditions of Latin astronomy. There were other, more broadly based criticisms of this tradition. Walcher himself criticized the accepted lunar computations (*vulgaris supputatio*) of the computists, which followed that nineteen-year cycle sanctioned by the authority of the church fathers and Bede.[70]

66. Menso Folkerts, "Adelard's version of Euclid's *Elements*," pp. 55–68 in Burnett, *Adelard of Bath*; Richard Lorch, "Some remarks on the Arabic-Latin Euclid," pp. 45–54 in Burnett, *Adelard of Bath*; Marshall Clagett, "The Medieval Latin Translations from the Arabic of the *Elements* of Euclid, with Special Emphasis on the Versions of Adelard of Bath," *Isis*, 44(1953):16–42.

67. Adelard of Bath, *De opere astrolapsus*; Poulle, "Le Traité de l'astrolabe d'Adélard de Bath."

68. North suggests that Adelard cast a group of horoscopes in the 1150s to determine the outcome of the conflict between King Stephen and Matilda. John D. North, "Some Norman Horoscopes," pp. 147–161 in Burnett, *Adelard of Bath*, at p. 159.

69. Adelard, *De opere astrolapsus*, pp. 158.5–12, 161.10–14, 216.1–6.

70. MS Auct. F.1.9, fol. 86va.

Peter Alfonsi rebuked his contemporaries for ignoring experience and limiting their knowledge of the heavens to what they had read in ancient authorities in the liberal arts like Macrobius.[71] Adelard of Bath praised the Arabs for following reason, unlike his contemporaries who credulously followed the mere appearance of authority like animals in a halter.[72] In France, Raymond of Marseilles attacked the inadequate astronomy of his Latin predecessors and contemporaries and critically compared new astronomical works, including a set of tables attributed to no less an authority than Ptolemy, against the "truth of the firmament itself."[73] Late in the century Daniel of Morley became disgusted with the bookishness, pretentious ignorance, and deliberate obscurity of the professors at Paris and left for Spain to learn astrology and cosmology, although not mathematical astronomy, from the "wiser philosophers of the universe," including the noted translator Gerard of Cremona.[74]

The dissatisfaction of these astronomers with both the traditional practice of computus and the teaching of the liberal arts combines an admiration for the new learning and techniques they had discovered in the Islamic world with a belief that they could apply these empirical and rational techniques to advance beyond their predecessors. The critical attitude of twelfth-century astronomers displays the ferment in the monastic and cathedral schools as they evolved to produce a new feature of medieval intellectual life: the university.

71. Millás Vallicrosa, "Pedro Alfonso," p. 99.

72. *Quaestiones naturales*, cap. 6, in Haskins, *Studies*, p. 40, n. 99; Ladner, "Terms and Ideas of Renewal," p. 7, nn. 36, 37; Beaujouan, "Transformation of the Quadrivium," p. 481.

73. Lemay, *Abu MaʿShar*, pp. 143–146; Haskins, *Studies*, pp. 96–98; Duhem, *Le Système du Monde*, vol. 3, pp. 205–207.

74. Haskins, *Studies*, pp. 126–127; Thorndike, *History of Magic*, vol. 2, pp. 171–179. Brian Stock places these critiques in the traditional tension between the obscurity of mythical cosmologies and the clarity of science and mathematics, but he largely ignores the appeal of mathematical astronomy's quantitative precision, which was lacking in cosmological discussions. Stock, *Myth and Science*, pp. 59–62. On Daniel's Greek and Arabic sources, see Silverstein, "Daniel of Morley," p. 180.

# CHAPTER TEN

# The rebirth of
# Ptolemaic astronomy

> Our purpose in this treatise is to describe the figure of the machine of the
> world, the center and figures of its constituent bodies, the motion of the
> higher bodies and the figures of their circles.
>
> Robert Grosseteste, *De sphaera*[1]

Western European scholars recovered the full content of geometrical astronomy
from the Islamic world in two stages. First, a group of scholars brought the
substance of this astronomy across the European linguistic frontier by seeking out
and translating from Greek and Arabic into Latin the mathematical and astronomical texts providing the theoretical basis of astronomy. These texts represented
substantial changes from the mere descriptions of the liberal arts tradition or the
calculating techniques presented in the *zījes*. The masters of the new universities
then completed the assimilation of this astronomy into the European intellectual
tradition as they applied it to traditional problems and taught and questioned its
cosmological implications in a torrent of original texts designed for their students.

## Translators and translations

Between 1060 and 1090 Sicily and southern Italy had been conquered by Norman
adventurers who established a kingdom with a trilingual Latin, Greek, and Arabic
literary tradition. The reconquest of Islamic Spain proceeded more slowly and
fitfully; Toledo fell in 1085 and Saragossa in 1118, but Córdoba and Seville were
not drawn back into the orbit of Christian Europe until 1236 and 1248, respec-
tively. Scholars and translators followed the expanding frontiers of Western Christendom, seeking not material wealth, but the scientific knowledge of the Islamic
world.[2]

Their interest in astronomy and astrology led twelfth-century translators to
begin with mathematical, astronomical, and astrological works.[3] These texts did
more than provide a theoretical basis for astronomical calculations. The astrolog-

---

1. Grosseteste, *De sphaera*, p. 11.
2. D'Alverny, "Translations and Translators"; LeMay, "Traductions de l'Arabe"; Haskins, *Studies*.
3. D'Alverny, "Translations and Translators," p. 451.

ical theory found in Abū Ma'Shar's *Greater Introduction* to astrology, translated by John of Seville around 1133 and again by Hermann of Carinthia in 1140, provided the first hints of Aristotelian natural philosophy's challenge to early medieval Platonism.[4]

An early expression of the more technical side of astronomy appeared in the 1140s as scholars modified astronomical tables to suit the meridians of specific European cities. But the production of astronomical tables did not require much knowledge of astronomical theory. The tables themselves had long been criticized for presenting techniques of calculation without either the reasons behind them or the proofs of their validity. In the tenth century Ibn al-Muthannā had criticized al-Khwārizmī's *zīj* on just these grounds, noting that "when books are composed in this way, they lack clarity and confound the reader."[5] He and other Arab scholars had sought to fill this gap with astronomical texts and commentaries.

Ibn al-Muthannā had presented the geometrical principles underlying a range of calculations in his detailed commentary on al-Khwārizmī's *zīj*. This commentary was translated into Latin by Hugh of Santalla for Bishop Michael of Tarazona, who occupied the see from 1119 to 1151.[6] Al-Farghānī's *On the Science of the Stars* provided a less technical introduction to the elements of geometric astronomy that went beyond the familiar discussions of the celestial circles and spheres. He introduced the fundamental problem of spherical astronomy, how the inclination of the zodiac and the latitude north or south of the equator influenced the time each sign took to rise, and also presented numerical values, based on Ptolemy's model of nested planetary spheres, for the dimensions of the universe. Despite these new concepts and data, al-Farghānī's presentation avoided most mathematical complexities. John of Seville first translated it in Spain around 1135; a second translation was made some thirty years later at Toledo by Gerard of Cremona.[7]

Although a desire to go beyond the practical techniques of the astronomical tables to a theoretical understanding of their underlying geometrical principles provided the immediate motive for many of these translations, the grail of the quest for astronomical learning remained the great work of that astronomer acclaimed by Cassiodorus and Isidore: Ptolemy's *Almagest*.

Both the Englishman Robert of Chester and the Slav Hermann of Carinthia studied geometry and astronomy in Spain. In the 1140s they planned a joint effort to translate the *Almagest* from Arabic, a project that was interrupted by other tasks, chief among them a translation of the Koran into Latin for the abbot of Cluny, Peter the Venerable, completed in 1143.[8] Although they abandoned their planned

---

4. LeMay, *Abu Ma'Shar*, pp. 9–19.

5. Ibn al-Muthannā, *Commentary*, pp. 150–151, cf. p. 15; Ibn al-Muthannā, *Comentario*, p. 95.

6. D'Alverny, "Translations and Translators," p. 443; Haskins, *Studies*, pp. 73–74.

7. Carmody, *Arabic Astronomical and Astrological Sciences*, pp. 113–116; al-Farghānī, *Rudimenta astronomica*; A. I. Sabra, "al-Farghānī," *Dictionary of Scientific Biography*, vol. 4, pp. 541–545; LeMay, *Abu Ma'Shar*, p. 13.

8. Haskins, *Studies*, pp. 120–121; d'Alverny, "Translations and Translators," pp. 429, 451.

translation of the *Almagest*, Robert and Hermann did continue with other translations of astronomical, astrological, and mathematical works.

This setback proved only temporary; the *Almagest* drew other translators and was soon translated twice into Latin. The earliest translation, made from the Greek around 1160, did not circulate widely. The translator was an unnamed student of medicine, who had traveled to Sicily with the express intent of translating the *Almagest*, having heard that a Greek manuscript of it had been presented by Emperor Manuel Comnenus as a gift to King Roger of Sicily.[9]

The same pattern lies behind Gerard of Cremona's much more influential translation. Gerard traveled to Toledo sometime before 1144, drawn by his love of the *Almagest*, which he could not find among the Latins. Besides the *Almagest*, his many translations include three crucial geometrical works: Euclid's *Elements* (building on the work of Adelard of Bath), Theodosius's *De sphaera*, and Menelaus's *Libri de figuris sphericis*.[10] With these translations the mathematical foundations of geometrical astronomy had been recovered. But Gerard did not limit himself to astronomical and mathematical works; his translations ranged widely over astronomy, astrology, mathematics, optics, medicine, and most of Aristotle's corpus of philosophical works.[11]

## Universities and the new learning

In time newly recovered learning transformed the university curriculum. To fully understand this change we would have to examine the place of Aristotelian philosophy in the medieval schools. Books have been written on this topic, which we can only touch in passing. Our primary concern is more limited: the assimilation of Greek and Arab geometrical astronomy by the Latin world and the discipline's incorporation into the curriculum of the new universities.

The universities grew in towns, nourished by the increasing prosperity of Western Europe, which shifted the focus of society from rural courts and monasteries to urban centers. In the world of learning this new focus was exemplified by a decline of monastic schools and the transformation of cathedral schools. Ultimately, a community of professional masters and their students gathered around the cathedral schools of a few important towns to create a new educational institution, the medieval university.

In astronomy, as in other disciplines, the masters prepared new texts for their

---

9.   Haskins, *Studies*, pp. 157–164, 191–193.

10.   Haskins, *Studies*, pp. 104–108; Richard LeMay, "Gerard of Cremona," *Dictionary of Scientific Biography*, vol. 15, pp. 173–192; Kibre, "*Quadrivium* in the Thirteenth Century," p. 185, n. 68; Kunitzsch, *Almagest*, pp. 83–87.

11.   The changing focus of Gerard's translations anticipates the evolution of scholarly interest from the practical arts of astronomy and astrology to the Aristotelian philosophy that would provide the core of the medieval university curriculum. LeMay, *Abu Ma'Shar*, pp. 350–352.

students. By the middle of the thirteenth century there were so many texts that Oliver of Brittany declared that "a day would scarcely suffice to completely tell of [astronomy's] innumerable books and authors."[12] What a contrast with the sixth century, when Cassiodorus could recommend only Varro and Ptolemy in his *Introduction to the Divine and Human Readings*.

The masters of theology reacted in various ways to the new philosophical and astronomical learning that their students brought from the arts to the study of theology. Some, such as Guiard of Laon, used astronomical concepts as familiar symbols to teach moral lessons. In his university sermon on the vigil of the Epiphany in 1231, he drew together the Magi who followed the star, the spherical model of the cosmos, and the customary connection of the order of monastic life to the order of the stars. He noted that the fixed stars, which do not move of their own volition but follow the motion of the firmament, provided proper models for monks, who should remain in their cloisters and follow the motion of the firmament, that is, of God. He bemoaned the fact that, in his time, willful monks wandered from place to place.[13]

A different view was expressed the next day by Peter of Bar-sur-Aube, who preached against applying the new learning to theology. He criticized his colleagues who employed philosophical opinions in their sermons and theological disputations, especially disparaging geometry and astronomy as the products of infidels, which were irrelevant to salvation and could even lead one into error and away from God. Despite the concerns of some theologians, masters and students of arts eagerly embraced the new astronomy.[14]

This interest in astronomy is not fully reflected in thirteenth-century descriptions of the university curriculum. These descriptions indicate that by the middle of the century Aristotelian natural philosophy had come to dominate the university curriculum. The liberal arts were increasingly seen as preparatory; the mathematical disciplines of the quadrivium are scarcely mentioned in university statutes. The principal astronomical work a student was required to read "pro forma" for the degree in arts was a treatise on the sphere (presumably that of Sacrobosco) and, at Oxford in the fourteenth century, the computus. Additional texts would make their way into the curriculum under other headings. Texts on the computus and *algorismus*, which treated calculations useful for astronomy, were

12. Oliverus Brito, *Philosophia*, Oxford, Corpus Christi College MS 243, fol. 4rb, cited in Lafleur, *Quatre Introductions*, p. 152, n. 113.

13. M.-M. Davy, *Les Sermons Universitaires Parisiens de 1230–1231*, Études de Philosophie Médiévale, 15 (Paris: J. Vrin, 1931), pp. 235–236.

14. Davy, *Les Sermons Universitaires*, p. 253. William of Auvergne, the bishop of Paris, contrasted philosophical knowledge of the order of the created heaven with true wisdom, which comes from knowing the order of the spiritual and sublime heaven where the author of the universe resides. Davy, p. 150. John of St. Giles directed scholars, who must answer for their works on the day of judgement, to the study of Scripture, although he admitted his own interest in how the world and the Sun were created. Davy, pp. 278–279.

sometimes taught as part of arithmetic, while texts dealing with measuring instruments such as the astrolabe and quadrant sometimes fell under the heading of
geometry.[15]


## The *corpus astronomicum*

The surviving texts make it clear that many masters and students went beyond
the minimum requirements to cover a full introduction to astronomy. They collected these related texts into a single unit, the *corpus astronomicum*, which covered
computus, the calendar, the nature of the celestial spheres, planetary theory, and
the uses of astronomical instruments. Some works in the *corpus* were selected from
the new translations, but most were works written in the beginning of the century
by Western European authors. Individual manuscripts of the *corpus* varied, reflecting a teacher's or student's choice among a range of competing works by different
authors, yet the similarity of topics indicates the overall uniformity of the astronomical curriculum. These collections reveal the scope and level of astronomical
teaching in the thirteenth century, just as the ninth-century Carolingian anthologies revealed the astronomical concerns of the Carolingian court.[16]

   The similarities and differences are striking. Computus is no longer central, but
it continues to be taught, using more critical texts written by authors who had
become familiar with the methods and conclusions of the new astronomy. Closely
related to these are calendars, containing much the same mixture of saints' days
with tabulations of the motion of the Sun through the zodiac, the length of
daylight, and the lunar phases found in earlier liturgical calendars, but drawing
on fresh theories and observations to provide more accurate values for these astronomical data. Manuscripts of the *corpus* commonly included the *algorismus*,
which discussed those arithmetical techniques useful for astronomical calculations.
Timekeeping by the Sun and stars was discussed in treatises on that now familiar
instrument, the astrolabe, and on the astronomical quadrant.

   Central among these new astronomical texts were a series of treatises going by
the generic title of *De sphaera*. Since treatises on the sphere only briefly mentioned
epicyclic planetary theory, by the end of the thirteenth century they were supplemented in most manuscripts by rudimentary geometrical descriptions of the
epicyclic models of planetary motion, the *Theorica planetarum*. The *Theorica planetarum* texts did not go deeply into the trigonometry needed to derive quantitative predictions from these descriptive models; the production of astronomical

15.   Kibre, "*Quadrivium* in the Thirteenth Century"; Delahaye, "La Place des Arts Libéraux,"
      pp. 168–170; Lafleur, *Quatre Introductions*, pp. 148–149; Weisheipl, "Curriculum at Oxford,"
      pp. 170–173.
16.   O. Pedersen, "*Corpus Astronomicum*." Optics, or *perspectiva*, is the other medieval example of the
      *scientiae mediae*.

tables, as opposed to their use, was a specialized matter that did not arise in the introductory course in astronomy. Even astronomical tables, however, ultimately became part of the curriculum, first appearing in the *corpus astronomicum* in the fourteenth century.

## The introduction to astronomy – De sphaera

At the core of the astronomical curriculum was a series of new textbooks on the sphere, modeled on the general pattern of al-Farghānī's *On the Science of the Stars* and replacing Martianus Capella, Macrobius, and the Carolingian anthologies with coherent introductions to the rudiments of spherical astronomy. Like earlier texts, treatises on the sphere provided general descriptions of the spheres and circles that formed the structure of the cosmos. But by drawing on the newly translated astronomy and geometry texts, they went beyond their predecessors to explain how this geometric structure produced subtle changes in observable phenomena. Yet none of these texts provided any instructions for calculations or ventured deeply into the complexities of planetary astronomy.

In this regard we should recall that few Western scholars ever advocated the study of astronomy as an end in itself. Astronomy, like the other liberal arts, had found its place in the educational system for its practical usefulness or for its propaedeutic value. Bede and Alcuin taught computus as a way to regulate religious ritual. Augustine saw an understanding of Creation as a key to the understanding of the Creator. Boethius had used astronomy to teach an ethical lesson: the immensity of the cosmos and the insignificance of earthly rewards. Even Ptolemy had justified mathematical astronomy as reforming human nature by accustoming its practitioners to "the constancy, order, symmetry and calm which are associated with the divine."[17] Most medieval masters valued astronomy for its contribution to the emerging study of philosophy.

If we sample a few treatises on the sphere, beginning with Robert Grosseteste's, written around 1215,[18] through the immensely successful *De sphaera* of John of Sacrobosco, who wrote around 1230,[19] to John Pecham's, written in the late 1270s, we see the authors increasingly using astronomy as a vehicle to discuss issues in physical cosmology.[20] This shift reflects the increasing role of Aristotelian natural philosophy in the medieval universities.

17. Ptolemy, *Almagest*, 1.1.

18. Grosseteste, *De sphaera*; Southern, *Robert Grosseteste*, pp. 142–146; James McEvoy, "The Chronology of Robert Grosseteste's Writings on Nature and Natural Philosophy," *Speculum*, 58(1983): 614–655, here pp. 617–618.

19. Thorndike, *Sphere of Sacrobosco*; O. Pedersen, "In Quest of Sacrobosco," pp. 190–192.

20. Robert Bruce MacLaren, "A Critical Edition and Translation, with Commentary, of John Pecham's *Tractatus de Sphera*," Ph.D. diss., University of Wisconsin – Madison, 1978.

    Another early treatise on the sphere is that of Campanus of Novarra, the author of a commentary on Euclid's *Geometry* and of several original astronomical works; Francis S. Benjamin,

Grosseteste's *De sphaera* reflects his belief that we cannot know natural philosophy without considering lines, angles, and figures. Consequently he emphasized the geometry of the heavens, presenting the model of what he, and his contemporaries, called the machine of the world with sufficient detail for his reader to visualize its overall geometrical structure. Although he presented the model of the heavens in clear and visually arresting language, he made few concessions to the geometrically faint of heart. "Let us imagine a great circle drawn through the aforesaid poles, and another, orthogonal to the first, drawn through the same poles, and these two circles are called colures. . . . Let us imagine again a great circle drawn . . . cutting the two aforesaid circles at right angles, and this is called the equinoctial [circle]."[21]

He seldom included either quantitative measurements or rigorous geometrical demonstrations in, what were in all other respects, precise geometrical descriptions of the motions of the heavens. Where Bede had explained the varying length of the day in terms of the sphericity of the earth, Grosseteste added the complex variation of the Sun's motion caused by the inclination of the zodiac (which he gave in the now standard degrees and minutes as 24° 33') and the eccentricity of the solar orbit.[22] Where Remigius of Auxerre had been puzzled by the Greek technical terms describing the Moon's inclined path, Grosseteste described it clearly and stated that the Moon must be within twelve degrees of the lunar nodes for a lunar eclipse to occur. His discussion of the Moon even included the moving eccentric from Ptolemy's lunar model.

Occasionally he went beyond geometrical demonstrations to consider physical arguments; he followed Aristotle's argument that since the earth is a heavy body, it would naturally tend towards the center to form a sphere. To this he added various *experimenta*, including both common experiences, as when we see the stars turning around the fixed pole star, and geometrical thought experiments, such as that people who travel north from Arim (a mythical city on the equator used as a prime meridian in many Arabic astronomical tables) will see the northern stars rise higher in the sky.[23]

Grosseteste's presentation seems ideally suited for a student who wants to learn the broad geometrical concepts governing the motion and visibility of the stars, Sun, and Moon – but not of the other planets. Like Bede, Grosseteste provided the theoretical basis for astronomical timekeeping and computus. Since problems of timekeeping and the calendar involve computation of the slow movement of the starry sphere, Grosseteste was obliged to judge between competing theories of its motion. He rejected Ptolemy's belief in the uniform motion of the starry

Jr., and G. J. Toomer, *Campanus of Novarra and Medieval Planetary Theory:* Theorica Planetarum (Madison: Univ. of Wisconsin Pr., 1971), pp. 13–14.

21. Grosseteste, *De sphaera*, pp. 13–14.

22. The corrections for the obliquity of the ecliptic and the solar eccentricity combine to produce the equation of time, a parameter commonly tabulated in astronomical tables.

23. Grosseteste, *De sphaera*, pp. 12–13.

sphere, in favor of the complicated theory of the oscillatory trepidation of the eighth sphere attributed to Thābit Ibn Qurrah. This section was certainly heavy going for an introduction to astronomy.[24]

If Grosseteste's uncompromising and somewhat dry introduction to the structure and motions of the *machina mundi* presents the theory behind basic astronomical computations, John of Sacrobosco's more appealing narrative reflects the interests of the typical student with less concern for technical details. Its dominance of the astronomical curriculum is indicated by its central place in the manuscripts of the *corpus astronomicum*, by the early appearance of scholastic commentaries on Sacrobosco's text, and by the hundreds of surviving manuscripts and over seventy early printed editions.[25] One early commentator hinted at the reason for Sacrobosco's appeal when he contrasted demonstrative astronomy in the tradition of Ptolemy and Thābit with narrative astronomy such as that of al-Farghānī, Martianus Capella, and Master John of Sacrobosco.[26]

Sacrobosco never reached the extremes of poetic allegory found in Martianus's *De nuptiis*, yet he did simplify his discussion of the structure of the heavens and interspersed it with discussions of the constellations, citations from the poets, and easy mnemonic verses like those used by Fulbert of Chartres. Sacrobosco's *Sphere* was a worthy successor to Martianus and the Carolingian anthologies, adding concepts from the new astronomy to their traditional concern with the constellations and the structure and dimensions of the world.

Like Grosseteste, Sacrobosco discussed the terrestrial and celestial spheres and the circles that gird the heavens: the equator, the tropics, the ecliptic, and the rest. But when Sacrobosco treated the motions of the ecliptic, he preferred to use the familiar names of the twelve signs that divide it, while Grosseteste had preferred to speak of angles and degrees. Sacrobosco, however, contrasted the risings of the signs presented by the astronomical poets such as Lucan, Ovid, and Virgil with the mathematical discussions of the rising of each arc of the zodiac found in astronomical and geometrical discussions of spherical astronomy. If one knows the astronomy, many of Sacrobosco's discussions seem repetitive and verbose. But for a beginning student, Sacrobosco's repeated examples and familiar names will drive home the orderly patterns of astronomical phenomena.

Sacrobosco's view of the place of astronomy in the arts curriculum is suggested by his treatment of the motions of the Moon. Where Grosseteste had provided a detailed description of the Ptolemaic model for the Moon, Sacrobosco briefly lumped together the model of the Moon with models of the other planets, ig-

---

24. Grosseteste, *De sphaera*, pp. 14, 25–27; Raymond Mercier argues convincingly that the theory of trepidation cannot be Thābit's. Mercier, "Astronomical Tables," pp. 105–106.

25. Thorndike, Sphere *of Sacrobosco*, pp. 25–40; O. Pedersen, "In Quest of Sacrobosco," pp. 183–184; O. Pedersen, *"Corpus Astronomicum,"* pp. 74–76. For the place of commentaries on Sacrobosco's *Sphere* in medieval cosmological discussions, see Grant, *Planets, Stars, and Orbs.*

26. Thorndike, Sphere *of Sacrobosco*, p. 413.

noring most of the substantial differences among them. His discussion of eclipses is similarly brief; he mentioned that solar eclipses cannot be seen everywhere but does not tell his reader why. Sacrobosco was not concerned with the niceties of technical astronomical detail. Yet his presentation did illustrate the order of nature, noting, as had Bede before him, that the solar eclipse at the Crucifixion must have been miraculous: "Either the God of nature suffers, or the *machina mundi* is dissolved."[27]

John Pecham tells us that he wrote his *De sphera* to teach beginners enough about the principal bodies of the world for them to understand the words of Sacred Scripture and to correct certain technical errors he found in Sacrobosco.[28] The principal error he noted was in Sacrobosco's discussion of the time it takes the various signs of the zodiac to rise. Citing geometrical demonstrations from Theodosius's *De sphaera*, Pecham charitably reinterpreted Sacrobosco's statement without naming him or attacking him outright.[29]

More significant, however, was Pecham's concern with the emerging tensions between mathematical astronomy and Aristotelian natural philosophy. Pecham proposed a scholastic *questio* to address a current philosophical issue in the arts faculty: whether astronomers' hypotheses (*positiones*) were really true. This *questio* raised the disagreements between the cosmology of concentric spheres presented by Aristotle's Arab expositor, al-Bitrūjī, and the "fables" of the astronomers, who held epicycles to be globes in the heavens carrying the stars in multiple contrary motions. Pecham could not fully resolve all the physical objections to the astronomers. Yet he came to the conclusion that while he had not yet heard how he could save all the appearances in the heavens while avoiding all philosophical inconveniences, the hypotheses of the mathematicians had greater probability.[30]

We can see two factors lying behind Pecham's ambivalent conclusion. The first was his own preference for combining physical and mathematical reasoning, an approach he employed effectively in his *Perspectiva communis*, a textbook as influential in optics as Sacrobosco's *De sphaera* was in astronomy. The second was a concern he shared with later commentators to reconcile the mathematical demonstrations of the astronomers with the physical reasoning of Aristotelian cosmology.

By 1250, the *De sphaera* was already required at Paris for the degree in arts.[31] But if we are to know how deeply masters took their students into this required text,

27. Sacrobosco, *Sphere*, pp. 114–117, 141–142.
28. Pecham, *Tractatus de sphera*, p. 88.1–8.
29. Pecham, *Tractatus de sphera*, pp. 128.3–132.2.
30. Pecham, *Tractatus de sphera*, pp. 138.10–144.4.
31. *De communibus artium liberalium*, Paris, BN MS Lat. 16390, fol. 198vb, cited in Lafleur, *Quatre introductions*, p. 131.

and how they approached the subject of astronomy, we need to look beyond the text itself to other sources. A series of guides to the study of the arts from the period 1240 to 1260 and a sampling of scholastic commentaries on the sphere indicate what was actually taught to, and expected of, students.[32]

These guides to the arts cite only ancient authorities, ignoring their contemporaries. Martianus Capella, who had provided the basic guide to astronomy before the reception of Arabic materials, was their principle authority on astronomy, although the influence of Sacrobosco's *De sphaera* is clearly visible. Hyginus, Aratus, and Ptolemy were occasionally mentioned, but specific texts were not cited; the only new texts referred to were those of Euclid and Aristotle.[33] The introductions suggest that the new Arabic astronomy had only scant influence on the introductory astronomical curriculum, a point confirmed by the masters' presentations found in commentaries on the sphere.

We are used to thinking of astronomy in terms of observation and calculation; we would expect a master expounding on the rising of the signs of the zodiac to move from the texts' qualitative discussions to a geometrical demonstration or a mathematical example. But such mathematical subtleties played only a minor role in their discussions. The typical medieval master chose to deal with other, albeit equally subtle, concerns flowing from Aristotle's philosophy.

Sacrobosco's introductory discussion of the nature of a sphere included a brief account of the four terrestrial elements and the fifth element, aether, of which the nine celestial spheres were made. His commentators took this opportunity to consider the geometrical definitions of a point, a line, a surface, and a sphere; the nature of the five elements and the relationships among them; how the spheres (or the planets they carry) exert an astrological influence upon the terrestrial region; and whether the celestial region was a single continuous whole, separated by mathematical spheres, or made up of contiguous distinct spheres.[34] These were all important issues in the Aristotelian natural philosophy of the schools, but they had little bearing on the observation or computation of astronomical phenomena.

From the sphere and its commentaries we might be led to conclude that, except for some new geometrical and philosophical concepts, the astronomy taught at Paris had scarcely changed in the three centuries from Remigius of Auxerre to John of Sacrobosco. But this would ignore the fact that neither Remigius nor his students had any other texts to supplement the simple geometric astronomy they had learned from late antique sources. In the thirteenth century, scholars could

---

32. Many commentators say they are writing for their students; one anonymous commentator tied his comments directly to the classroom, saying "These are what I set forth in the schools concerning the sphere and wrote in honor of our Lord Jesus Christ. . . ." Thorndike, *Sphere of Sacrobosco*, p. 58.

33. *Accessus philosophorum* (1230 × 1240), *Compendium circa quadrivium* (post-1240), *Philosophia disciplina* (ca. 1245), and *Divisio scientiarum* (1250 × 1260); ed. Lafleur. *Quatre introductions*.

34. Robertus Anglicus, *On The Sphere*, pp. 143–157; Michael Scot, *On The Sphere*, pp. 255–284.

supplement their reading of the sphere with the rest of the *corpus astronomicum* and with a wide range of other astronomical and mathematical texts by Greek, Arab, and Latin authors.

### Computus and the calendar

While the introductory astronomy of the sphere was augmented by the new astronomical theory, computus texts and calendars were confronted with new astronomical data and theory found in the translations. Whereas Walcher of Malvern could only note those cases where observed eclipses disagreed with his improved lunar tables, astronomers now had theoretical demonstrations that the traditional calendar erred both on the dates of the solstices and equinoxes and on the cyclical motion of the Moon. As early as 1175 an anonymous English computist was citing Greek and Arabic sources as evidence for the correct date of the vernal equinox.[35]

In the many new texts on computus, and in the calendars derived from them, we see scholars critically examining traditional methods of calculating the date of Easter in the light of new astronomical data. Robert Grosseteste returned several times to this problem.[36]

In his early *Compotus I*, Grosseteste expressed his awareness of the problems of the ecclesiastical calendar and his intent to consider them in greater detail later. The problem that drew his attention was the familiar one of the changing dates of the solstices and equinoxes. He focused less on the significance of the equinoxes for Easter, than on the relationship of the solstices to the birthdays of Christ and of John the Baptist, which were supposed to coincide with the turnings of the Sun. Yet Grosseteste already knew that the Julian year differs from the solar year by one day in 120 years, from which he determined that in the past 1200 years Christmas had retrogressed from the winter solstice by ten days.[37]

Some twenty-five years later, when Grosseteste reconsidered this problem in his *Compotus correctorius*, he used parameters from the new Arabic astronomy, the computational techniques of sexagesimal fractions, and the simplest kind of observations to examine the calendar with greater confidence and precision. Beginning with the question of the length of the solar year, he compared the value of Hipparchus and the computists (365 ¼ days) with that of Ptolemy (365 + ¼ − 1/300 day) and that of al-Battānī (365 + ¼ − 1/100 day). He concluded that al-Battānī's result agrees most closely with "the experience of our time" (*experimen-*

35.   Oxford, Bodleian Library, MS Auct. F.1.9, ff. 95vb–96ra; Haskins, *Studies*, pp. 87–88, citing an anonymous computus in British Library, MS Cotton Vitellius A. xii, ff. 87–97v.

36.   Southern, *Robert Grosseteste*, pp. 125–131, 139–140, 142–146; Dales, "Computistical Works Ascribed to Grosseteste."

37.   Dales, "Computistical Works Ascribed to Grosseteste." Dales sees no reason to accept Jennifer Moreton's doubts about Grosseteste's authorship. See note 47.

*tum nostri temporis*), "since according to the scriptures, our Lord Jesus Christ was born at the winter solstice, which now precedes the birthday of the Lord by about as many days as centuries have passed since the time of his birth."[38]

Turning from the Sun to the Moon, Grosseteste computed the error of the 19-year lunar cycle using Ptolemy and Hipparchus's value of the mean length of the lunar month (29;31,50,8,9,20 days), and al-Zarqālī's truncated value of 29;31,50 days. Examining the 76-year lunar cycle, Grosseteste noted that in four such cycles, or 304 years, the error accumulates to a bit more than a day using al-Zarqālī's value, a bit less than a day using Ptolemy's.[39] To those who might claim that the computists were correct, while Ptolemy and al-Zarqālī were wrong, Grosseteste simply pointed out that the astronomers' "tables do not mislead us in any perceptible way as to the hour of eclipses,"[40] which would not be the case if the astronomers' value for the length of the lunar month was incorrect.

Both Grosseteste and his compatriot Roger Bacon suggested that the Arabic calendar offered the key to a simple reckoning of the new Moon. Thirty such Arabic years contain 360 lunar months, which (using al-Zarqālī's value for the length of the month) precisely equal 10631 days.[41]

Roger Bacon used a richer range of sources and treated the problem in finer detail than had Grosseteste. Where Grosseteste had claimed that the solstitial feasts of Christmas and the birth of John the Baptist were attested in Scripture, Bacon more properly attributed this tradition to later scriptural commentators. His discussion of the changing date of the solstices and equinoxes considered the unequal length of the quarters of the year, using an eccentric model of solar motion.[42] His discussion of the new Moon included many complications introduced by the new astronomy: the nonuniform motion of the Sun and Moon and the various factors that influence the first visibility of the new Moon on the horizon.[43] Bacon noted, following Ptolemy's *Almagest*, that a cycle in which the periods of all these diversities of motion would be reconciled is "344 calendar years, plus 361 days and one hour." Bacon emphasized the variations of the motion of the Sun and Moon and set out to compute the longest and shortest lunar months reckoned from conjunction to conjunction; the task was too complex, and his text leaves blanks where the exact number of hours was to be inserted.[44]

38. Grosseteste, *Compotus*, p. 215.
39. Grosseteste, *Compotus*, pp. 232–236; Baur, *Philosophie des Grosseteste*, pp. 53–55; North, "Western Calendar."
40. Grosseteste, *Compotus*, p. 235.
41. Grosseteste, *Compotus*, pp. 232–233, 237.
42. Bacon, *Compotus*, 1.7, pp. 29–37.
43. Bacon, *Compotus*, 1.15, pp. 64–66.
44. Bacon, *Compotus*, 1.12–13, pp. 50–58. Bacon uses a value for the length of the lunar month of 29;31,50,8,9,9 days, which differs from Ptolemy's in the last sexagesimal place.

John of Sacrobosco also included a *Compotus ecclesiasticus* among his many astronomical texts. Again Ptolemy plays a role, but Sacrobosco's discussion of the problem of the solstices and the equinoxes reveals that he and his students were more familiar with the numerical data found in astronomical tables than with the *Almagest* itself. He ignored Ptolemy's explicit statement that the tropical year differed from the Julian value by one day in 300 years and instead derived from the tables a difference of one day in 288 years.[45] Unlike Grosseteste, Sacrobosco accepted Ptolemy's value as correct, and he concluded that the ancients at the time of Christ had erred by six days in determining the day of the solstices on which Christmas and St. John's Day were celebrated.[46]

The theoretical questions raised in the new texts on computus took some time to work their way into astronomical calendars. These new calendars followed the pattern set by Walcher of Malvern, tabulating the time of every mean conjunction of the Sun and Moon over four nineteen-year lunar cycles.

One calendar covering the seventy-six years from January 1208 through December 1283, attributed in some manuscripts to Robert Grosseteste, gives the time of new Moon to the nearest whole hour, reckoned (according to one manuscript) for the meridian of Paris. Yet it is no more accurate than Walcher of Malvern's calendar of a century earlier.[47] The computed times of new moons differ sufficiently from Walcher's to show that the calendar's author performed his own calculations, but we cannot tell whether his starting point was an astronomical observation like Walcher's or a new Moon calculated from the new astronomical theory. Like Walcher, however, he based his further computations on the traditional nineteen-year lunar cycle rather than on the more accurate values of lunar motion available from Arabic sources.

These more accurate values were first used in the calendars of William of St. Cloud and Peter Nightingale, written in the 1290s. Although sharing the same elements, these two calendars differ both in content and in style. Peter's calendar is a working calendar, apparently written for the university classroom, consisting of tables for the twelve months, preceded by a brief set of instructions, or *canones*, occupying less than two pages. William's calendar was written for Queen Marie of Brabant, the widow of Philip III, with a lengthy preface on the value of astronomy and extensive discussions of the applications of the calendar.[48]

45. O. Pedersen, "In Quest of Sacrobosco," p. 209; Ptolemy, *Almagest*, 3.1.

46. O. Pedersen, "In Quest of Sacrobosco," p. 210.

47. Grosseteste, "Die Neumondtafel." I have not been able to locate the study by Jennifer Moreton, "Robert Grosseteste, John of Sacrobosco, and the Calendar," presented at the Warburg Institute Grosseteste Symposium in May 1987. Moreton questions Grosseteste's authorship of the *Kalendarium*, considering it nearly identical to that of Roger of Hereford. Dales, "Computistical Works Ascribed to Grosseteste," pp. 74–5, 78.

48. Peter Nightingale, *Opera quadrivialia*; William of St. Cloud, *Kalendarium Regine*; W. E. van Wijk,

Their layout is traditional, giving one or two pages for each month and a line for each day, with notations for saints' days and other fixed religious feasts; the dates of the solstices, equinoxes, and changes of the seasons; the days on which the Sun entered each sign of the zodiac; the day and hour of every computed mean new Moon in the 76-year cycle; and the so-called golden number, which indicates the date of the ecclesiastical new Moon in the 19-year cycle. In addition, many copies add the geometrically computed height of the Sun at noon and the length of day and night.[49]

Although following the layout of earlier calendars, William and Peter abandoned the computists' strict adherence to the 19-year luni-solar cycle, with its assumption that a conjunction would occur on the same date and time after 76 years. These new tables employed Hipparchus's value for the mean length of the month of 29 days, 12 hours, 44 minutes, 3 seconds, and 20 "thirds" of an hour, which could be found in Ptolemy, al-Khwārizmī, or the *Tables of Toledo*. Peter computed the mean times of conjunctions using the *Tables of Toledo*, while William adjusted his computations by 40 minutes on the basis of his own eclipse observations.[50]

This increased precision had its drawbacks: more precise calendars were no longer perpetual. A full 76-year cycle of 940 such lunations is 6 hours, 40 minutes short of the computists' value of 27,759 days. The calendars needed to be adjusted by about six hours to be valid for the next 76-year cycle. Besides these practical limits, neither William nor Peter applied epicyclic models to compute the true time of conjunctions; their calendars tabulated only mean conjunctions of the Sun and Moon. Computing the true time of 940 conjunctions using the *Tables of Toledo* was just too much work.

Like his predecessors, William was concerned that the discrepancy between the astronomically correct values and the date of religious feasts computed according to traditional methods made the Church appear ridiculous to Jews and Muslims. Furthermore, he recognized that although his calendar would not be used at present to compute religious feasts, it was especially valuable for physicians, who need to know the phases of the Moon and its place in the zodiac.[51]

The new calendars also give the astronomically correct day for the Sun's entry into each sign, the geometrically computed height of the Sun at noon, and the lengths of day and night. These new calculations of the length of daylight superseded the traditional simple approximations found in monastic customaries and ecclesiastical calendars since Gregory of Tours and Bede of Jarrow. William noted that by using these figures to regulate a *horologium*, his calendar could help deter-

*Le Nombre d'Or: Étude de Chronologie Technique* . . . (The Hague: Martinus Nijhoff, 1936), pp. 29–31, 38–41.

49. Peter Nightingale, *Opera quadrivialia.*, pp. 331–337; William of St. Cloud, *Kalendarium Regine*, pp. 96–97, 202–203.

50. Peter Nightingale, *Opera quadrivialia*, pp. 59–61; Harper, "The *Kalendarium Regine*," pp. 46–48.

51. William of St. Cloud, *Kalendarium Regine*, pp. 112, 119, 213, 217.

mine the correct time of prayers by day and night, and especially just before dawn.[52]

### Observing the heavens – instrument texts

Astronomical instruments provided a more direct way to determine time, which was discussed in the *corpus astronomicum*. The familiar astrolabe continued to be taught, chiefly using the late thirteenth-century treatise on the construction and use of the astrolabe falsely attributed to Māshāʾallāh. Pseudo-Māshāʾallāh presented all the usual problems concerned with measuring the positions of the stars and the Sun, finding the time of day, determining one's latitude, and orienting oneself. But pseudo-Māshāʾallāh dealt with two new categories of problems. He told how to determine the position of the planets and find whether the motion of the planets is direct or retrograde, suggesting a new element of astronomical interest. Other chapters treat astrological concerns, showing how to use the astrolabe to divide the heavens into the twelve astrological houses and explaining the aspects of the planets.[53]

The astrolabe was, however, a complex and costly instrument. A simpler, if more limited, instrument was the old quadrant. This Arabic invention was de-scribed in a group of treatises on the *Quadrans vetus* which soon entered the *corpus astronomicum*. The most commonly used text was written at Montpellier sometime between 1250 and 1284 by a John (or Robert) the Englishman, while among the authors of less widely used texts we find John of Sacrobosco and Campanus of Novara.[54] The quadrant's main astronomical use was for finding the time of day by observing the height of the Sun. On its outer arc was a sliding cursor with a calendar scale for calculating the noontime height of the Sun on any day of the year; the texts often supplemented this with tables for finding the noontime height of the Sun for any day without using the cursor. Once the quadrant was set for the Sun's noontime altitude, a single observation of the current height of the Sun would give the time of day in unequal, that is, seasonal, hours.

52. William of St. Cloud, *Kalendarium Regine*, pp. 102, 206.
53. [Ps.-] Messahalla, "De compositione astrolabii; De operatione vel utilitate astrolabii," pp. 295–232 in R. T. Gunther, *Early Science in Oxford*, vol. 5 (Oxford: Oxford Univ. Pr., 1929).
    Māshāʾallāh (d. ca. 815) was not the author of this treatise. It has been shown to be a late thirteenth-century Western compilation from various tenth- and eleventh-century Spanish Ar-abic sources. Paul Kunitzsch, "On the Authenticity of the Treatise on the Composition and Use of the Astrolabe Ascribed to Messahalla," *Archives Internationales d'Histoire des Sciences*, 31(1981): 42–62.
54. Nan L. Hahn, *Medieval Mensuration:* Quadrans Vetus *and* Geometrie Due Sunt Partes Principales . . . , *Transactions of the American Philosophical Society*, vol. 72, pt. 8 (Philadelphia, 1982), pp. xv–xxi, xxxii–xxxiii; O. Pedersen, *"Corpus Astronomicum,"* p. 75; O. Pedersen, "In Quest of Sac-robosco," pp. 185–186; Benjamin and Toomer, *Campanus of Novara*, p. 15.

### *The theory of the planets* – Theorica planetarum

Geometrical descriptions of the motions of the five planets were restored to the curriculum, from which they had been absent since late antiquity, only with the appearance in the mid-thirteenth century of several texts called *Theorica planetarum*. The usual text is an anonymous introduction to Ptolemaic planetary theory, defining technical terms and describing the geometrical models for the motions of the Sun, Moon, and other planets, including the especially complex model for the planet Mercury.[55]

Besides explaining the models, the text uses them to provide clear geometrical explanations of the computational elements found in Ptolemaic astronomical tables. The concluding section provides concise instructions for computing the positions of the heavenly bodies using the tables (although confusing Ptolemaic and Hindu procedures when discussing planetary latitudes), tells how to modify astronomical tables to suit the longitude of other places, and gives rules for rudimentary astrological calculations.[56]

A much lengthier *theorica planetarum* was written between 1261 and 1264 by Campanus of Novara, a cleric associated with the papal court. Campanus's text was not incorporated into the *corpus astronomicum*, probably because of its greater length and detail. Campanus covered all the major points of Ptolemaic theory, omitting only a discussion of planetary latitudes. In addition, he incorporated two elements lacking in the more usual text.[57]

One is the discussion of the distance of the planets according to the model of Ptolemy's *Planetary Hypotheses*, which Campanus took from al-Farghānī's introduction to astronomy. Campanus clearly thought in terms of Ptolemy's model of physically real nesting spheres, giving their distances from the earth in both earth radii and miles. He found the outermost surface of the sphere of Saturn, and hence the inner surface of the sphere of the fixed stars, to be 73,387,747 100/660 miles from the earth and the area of that sphere to be 67,706,715,144,825,054 38/99 square miles.[58]

Campanus's other innovation provided instructions for making and using an astronomical instrument, the equatorium, which he apparently knew from an unknown Arabic source. The equatorium is a scale model of the Ptolemaic system for a planet, designed as a mechanical computer for converting the mean motions of the planet, found in the tables, to the planet's true position without extensive computations. Although no instruments meeting Campanus's exact description survive, he was influential in stimulating the later development of more compact

55. O. Pedersen, "*Theorica Planetarum* Literature"; O. Pedersen, "*Corpus Astronomicum*," pp. 76–77.
56. Anonymous, "Theory of the Planets."
57. Campanus of Novara, *Theorica planetarum*.
58. Benjamin and Toomer, *Campanus of Novara*, pp. 34, 53–56, 342–345, 356–363.

and sophisticated versions of the device. These made their appearance in the *corpus astronomicum* in the next century, providing a simpler way to compute the positions of the planets.[59]

By the end of the thirteenth century, astronomy had found a secure, if limited, home in the university curriculum. Astronomy was less central in the scholastic curriculum, dominated as it was by Aristotelian philosophy, than computus had been for Bede, Alcuin, and Hrabanus. For most students in the arts faculty the study of astronomy went little beyond the study of the sphere, usually given in Sacrobosco's popular presentation. But even Sacrobosco's elementary discussion of the geometrical structure of the heavens surpassed former treatments. If university statutes and surviving texts have any meaning, most philosophers, clerics, lawyers, and physicians learned the rudimentary spherical astronomy of Sacrobosco's *De sphaera*.

The surviving manuscripts of the *corpus astronomicum* show that the universities also provided opportunities for masters and students interested in the more advanced study of astronomy. And there were motives for studying more advanced material. The new developments in astronomical timekeeping and computus were of use to those in the arts and theology faculties who would make their careers in the Church. William of St. Cloud reminds us that the physicians, who studied in the medical faculty, needed to know astronomy in order to treat their patients properly.[60] And some, as always, studied astronomy for its own intrinsic interest.

### Astronomy outside the universities

Although astronomy was now taught in the universities, many astronomers continued the time-honored practice of finding patrons at court. In the 1290s when William of St. Cloud dedicated his treatises to royal patrons, he echoed the familiar theme of the philosopher-king, asking that these modern times revive the ways of antiquity when kings and princes esteemed the arts, loved learning, and honored scholars so that armies would prosper through the flowering of learning.[61] He sketched the association of learning and political dominion from the Chaldeans, through Aristotle and his student Alexander, to Charlemagne, king of the Franks and emperor of the Romans, who promoted studies and "always carried scientific books with him and had scientific discussions by day and night with the astronomers and other experts who accompanied him." He closed with Boethius's

---

59.  Benjamin and Toomer, *Campanus of Novara*, pp. 30–33; O. Pedersen, *"Corpus Astronomicum,"* pp. 79–80.
60.  William of St. Cloud, *Kalendarium Regine*, pp. 119, 217.
61.  Harper, "The *Kalendarium Regine*"; Mancha, "Astronomical Use of Pinhole Images"; Duhem, *Le Système du Monde*, vol. 4, pp. 10–19.

maxim from Plato, that government would flourish when philosophers governed or rulers studied philosophy.[62]

If William's plea for patronage makes him sound like a practicing astronomer, the series of astronomical observations he carried out between 1285 and 1292 reinforce that image. William was aware of the disagreements among ancient and Arabic astronomers, and between 1285 and 1292 he observed the Sun, Moon, Mars, Jupiter, and Saturn to test the reliability of astronomical tables and improve their accuracy.

William observed the noontime height of the Sun at the summer and winter solstices and at the vernal equinox. From these three observations he computed the maximum declination of the Sun (i.e., the obliquity of the ecliptic) as 23° 34', within two minutes of the modern value for that era. He estimated that the true time of the equinox was sixteen hours after his observation at noon on 12 March 1290, that is, at 4:00 in the morning on 13 March, within six hours of the value from modern tables and a great improvement over the conventional values in ecclesiastical calendars.[63] But he did not just question the traditional date of the equinoxes; he used these observations to resolve differences among the new authorities. Was the slow motion of the eighth starry sphere uniform, as Ptolemy had maintained, or was there an oscillatory trepidation in this motion? William's own observations led him to reject both the theory of trepidation, attributed to Thābit Ibn Qurrah, and Ptolemy's estimate that the stars moved one degree in somewhat less than a century, accepting instead al-Battānī's value of one degree in sixty-six years.[64]

Like his predecessors, William also turned to the problem of the motion of the Moon, using observations of eclipses to refine the values of the Moon's mean motion (*medius motus*). Noting that computed mean conjunctions always preceded the observed time of the eclipse, William concluded that the computed time of mean conjunctions needed to be corrected by about forty minutes. William did not propose any changes to the *argumentum*, which accounts for the nonuniformity of the Moon's motion, although he was aware that this also needed revision.[65]

In his *Almanach planetarum*, which deals with the motions of the seven planets, he described a new technique to measure the apparent diameter of the Sun and the time and magnitude of solar eclipses by projecting the Sun's image through a small aperture onto a distant surface. He realized that observations of the changing apparent diameter of the Sun through the year could yield a direct measurement of the eccentricity of the solar orbit. William's proposal of direct

62. Harper, "The *Kalendarium Regine*," pp. 78–80, 186–188.
63. Harper, "The *Kalendarium Regine*," pp. 42–43. The value of the obliquity of the ecliptic for 1290 given by the formula of Bretagnon and Simon is 23° 31' 53". Tuckerman's tables place the equinox at 22:34 Paris mean time on 12 March 1290. Bretagnon and Simon, *Tables*, p. 6; Tuckerman, *Planetary Positions*, vol. 59.
64. Harper, "The *Kalendarium Regine*," pp. 43–45.
65. Harper, "The *Kalendarium Regine*," pp. 46–48.

measurement of the Sun's changing distance implies that, unlike some contemporary Aristotelian natural philosophers, this working astronomer did not consider eccentrics and epicycles to be "fables" used to compute the changing position of the Sun. They were real and had observable effects on the Sun's distance as well as on its position.[66]

William did not carry out his proposed observations of the solar diameter, but he did observe conjunctions of Saturn with Jupiter and of Mars with the Moon and with a star in the constellation Scorpio. He then used these observations to propose small corrections to the values of the mean motions given in the *Tables of Toulouse* and incorporated these corrected values into his planetary almanac, which gave the positions of the planets for twenty years beginning with the year 1292.[67]

## The legacy of early medieval astronomies

The emergence of a wide range of original astronomical texts in the thirteenth century represents the culmination of that long process by which Latin scholars recovered the astronomy of Ptolemy and his Arab successors and provided it with a secure social and intellectual environment in the medieval universities. This process reveals how the traditions of early medieval astronomy, securely rooted in the fertile soil of early medieval culture, had provided contexts within which ancient astronomy came to be seen as having practical uses, as well as philosophical significance.

The tradition of solar calendars connected with local rituals had lost its connection with solar horizon observations by the ninth century, when Alcuin of York's advocacy of the feast of All Saints provided the last sign of a formal awareness of this astronomical tradition. Yet discussions of computus from Bede to Helperic continued to describe observations of the Sun near the horizon, and diagrams in computistical manuscripts marked the changing places of sunrise and sunset at the births of Christ and of John the Baptist and at the equinoxes.

The other ritual calendric tradition, Easter computus, played a more significant role in the development of medieval astronomy. From the Irish computists and Bede to Alcuin and Hrabanus, computing the single, universally observed, date of Easter, when light triumphed over darkness, had been the central element of astronomical instruction. And even as the ancient tradition of the astronomy of

---

66. Harper, "The *Kalendarium Regine*," pp. 54–58; Mancha, "Astronomical Use of Pinhole Images," pp. 279–283; Duhem, *Le Système du Monde*, vol.4, pp. 17–18. It should be stressed here that neither the eccentricity of the solar orbit nor the size of the solar epicycle were directly measured quantities in Ptolemaic astronomy; they were inferred from the Sun's nonuniform annual motion. Hence philosophers could, and did, dismiss eccentrics and epicycles as fictional predicting devices.

67. Harper, "The *Kalendarium Regine*," pp. 48–51, 65–68, 40–41.

the liberal arts typified by Macrobius began to play a greater role in instruction, the astronomical problem of determining the vernal equinox and the subsequent Paschal Full Moon remained important.

The astrolabe, which had already transformed the local observations of stars at night that regulated monastic prayers, incorporated a quantitative model of the heavens, including new data on solar motions, and made possible precise investigations of the lunar calendar. Observations of eclipses soon led Walcher of Malvern and his colleagues to realize that an improved lunar calendar required the subtle astronomical techniques known to the Arabs.

These complex interactions of early medieval astronomies stimulated the quest for Arabic learning that dominated the twelfth century. In turn, that new learning so transformed these traditions that they came to depend increasingly on the geometrical astronomy taught in the universities.

By the thirteenth century, the tradition of solar horizon observations to determine the course of the year had all but vanished. Consideration of the rising Sun retained its religious symbolism and could be explained in terms of geometrical astronomy but, unlike the other practical traditions, contributed nothing further to astronomical study.[68]

The tradition of monastic timekeeping had been transformed by new astronomical instruments and texts describing their construction and use. For a while monastic texts continued to call for watching the stars to determine the times of prayer, but before the end of the century mechanical clocks would begin to replace the astrolabe as the regulator of liturgical time.[69] Astronomical observations came to regulate the clock, which became a mechanical image of the heavens.

The computistical tradition employed the new astronomy to clarify difficulties that had long troubled computists. Texts on computus and the calendar proposed a variety of changes in traditional practice that would produce an acceptable calendar. Although the reform of the calendar they called for would not come until the sixteenth century, the ecclesiastical calendar did provide thirteenth-century scholars with an ongoing focus for discussions of astronomical theory.

It is the geometrical astronomy of the liberal arts that was transformed most fundamentally. With the recovery of its full content and mathematical rigor, geometrical astronomy reclaimed its central place in astronomical study. While texts on the sphere adhered closely to the ancient cosmological tradition epitomized by Martianus Capella, they went beyond a limited understanding of the changing

68. In discussing how to find true east, where the Sun rises on the equinox, William of St. Cloud noted that "churches should be founded along this line, since the true east signifies our Lord Jesus Christ." William of St. Cloud, *Kalendarium Regine*, pp. 108, 210.

69. A series of mechanical clocks were made for English cathedrals and monasteries, beginning with one constructed by the Austin Canons of Dunstable Priory in 1283. C. F. C. Beeson, *English Church Clocks, 1280–1850: History and Classification* (London: Antiquarian Horological Society, 1971), pp. 13 ff.

length of daylight to a full comprehension of the subtle influences of the geometrical structure of circles and spheres on the seasonal changes of the times of
sunrise and sunset and rudimentary introductions to the motions of the planets.
For those interested in the more advanced study of astronomy, texts on the sphere
were supplemented by the *corpus astronomicum* and the works of Ptolemy and his
successors.

Latin astronomers had sufficiently assimilated the tradition of Ptolemaic geometrical astronomy that they could now conduct the kind of independent observations needed to propose corrections to both the traditional ecclesiastical calendar
and the new Ptolemaic astronomical tables. Through their mastery of Ptolemaic
astronomy and its incorporation into the university curriculum, medieval astronomers had endowed their successors with a mature academic specialty. This quantitative geometrical understanding of the machine of the world and its parts took
its substance from the Greek and Arabic traditions of geometrical astronomy,
grafted onto the hardy rootstock of early medieval astronomies.

# Bibliography

## Sources

Abbo of Fleury. *De ratione spere.* In "Two Astronomical Tractates." Dordrecht, 1985.

Abbo of Fleury. "Further Astronomical Material of Abbo of Fleury." Ed. R. B. Thomson. *Mediaeval Studies,* 50(1988):671–673.

Abbo of Fleury. "Two Astronomical Tractates of Abbo of Fleury," Ed. R. B. Thomson. Pp. 113–133 in J. D. North and J. J. Roche, *The Light of Nature.* Dordrecht, 1985.

Adelard of Bath. *De opere astrolapsus.* In B. G. Dickey, "Adelard of Bath." Toronto, 1983.

*Admonitio generalis.* Ed. A Boretius. *MGH,* Leges., Cap., I, 52–62. Hannover, 1883.

Adomnan. *Adomnan's Life of Columba.* Ed. and trans. A. O. Anderson and M. O. Anderson. London, 1961.

Aelfric. *Ælfric's Lives of the Saints.* EETS, 94. London, 1890.

Alcuin of York. *The Bishops, Kings, and Saints of York.* Trans. Peter Godman. Oxford, 1982.

Alcuin of York. *Carminae.* Ed. E. Duemmler. *MGH,* Poet. 1. Berlin, 1880.

Alcuin of York. *Epistolae.* Ed. E. Duemmler. *MGH,* Epp. 4, 1–493. Berlin, 1895.

Aldhelm. *Epistulae.* Ed. R. Ehwald. *MGH,* Auct. Antiq. 15. Berlin, 1919.

pseudo-Anatolius. *De ratione paschali.* In B. Krusch, *84jährige Ostercyclus.* Leipzig, 1880.

Angelomus of Luxeuil. *Enarrationes in libros Regum.* PL 115, 243–552. Paris, 1881.

Anonymous. "The Theory of the Planets." Trans. O. Pedersen. In E. Grant, *A Source Book in Medieval Science.* Cambridge, Mass., 1974.

Ascelinus. [*Quomodo fiat astrolabium*]. In W. Bergmann, *Innovationen im Quadrivium.* Stuttgart, 1985.

The Astronomer. *Vita Ludovici Imperatoris.* PL 104, 927–980. Paris, 1864.

Audoenus of Rouen. *Vitae Eligii episcopi Noviomagensis liber II.* Ed. B. Krusch. *MGH,* Scr. Rer. Merov. 4. Hannover and Leipzig, 1902.

Augustine. *De civitate Dei.* CCSL 47–48. Turnholt, 1955.

Augustine. *De divinatione dæmonum.* PL 40, 581–592. Paris, 1887.

Augustine. *De Genesi ad litteram.* Ed. J. Zycha. CSEL 28.1. Vienna, Prague, and Leipzig, 1894.

Augustine. *Enarrationes in Psalmos.* Ed. E. Dekkers and J. Fraipont. CCSL 40. Turnholt, 1956.

Augustine. *Quaestionum in Heptateuchem libri VII.* Ed. J. Zycha. CSEL 28.2. Vienna, Prague, and Leipzig, 1895.

Augustine. *Sermones.* PL 38. Paris, 1863.

pseudo-Augustine. *De mirabilibus sacrae scripturae.* PL 35, 2149–2200. Paris, 1864.

pseudo-Augustine. *Questiones veteris et novi testamenti CXXVII*. Ed. A. Souter. *CSEL* 50. Vienna and Leipzig, 1908.

pseudo-Augustine. *Sermo in Pascha*, 164. *PL* 39, 2066–2068. Paris, 1865.

Ausonius. *Commemoratio professorum Burdicalensium*. Trans. H. G. E. White. Loeb. Cambridge, Mass., 1919.

Bacon, Roger. *Compotus fratris Rogeri Baconis*. Ed. R. Steele. *Opera hactenus inedita Rogeri Baconi*, Fasc. 6. Oxford, 1926.

Bede. *Bedae opera de temporibus*. Ed. C. W. Jones. Cambridge, Mass., 1943.

Bede. *De temporibus liber*. Ed. C. W. Jones. *CCSL* 123 C. Turnholt, 1980. Also in *Bedae opera de temporibus*. Cambridge, Mass., 1943.

Bede. *De temporum ratione*. Ed. C. W. Jones. *CCSL* 123 B. Turnholt, 1977. Also in *Bedae opera de temporibus*. Cambridge, Mass., 1943.

Bede. *Historia ecclesiastica gentis Anglorum*. Ed. B. Colgrave and R. A. B. Mynors. Oxford Medieval Texts. Oxford, 1969. *A History of the English Church and People*. Trans. L. Sherley-Price. Baltimore, 1958.

Bede. *In Regum librum XXX quaestiones*. Ed. D. Hurst. *CCSL* 119. Turnholt, 1962.

Bede. *Libri quatuor in principium Genesis*. Ed. C. W. Jones. *CCSL* 118 A. Turnholt, 1962.

pseudo-Bede. *De argumentis lunae*. *PL* 90, 701–728. Paris, 1862.

pseudo-Bede. *De mundi celestis terrestrisque constitutione: A Treatise on the Universe and the Soul*. Ed. C. Burnett. Warburg Institute Studies and Texts, 10. London, 1985.

pseudo-Bede. *Homiliae subditae*. *PL* 94, 267–516. Paris, 1862.

pseudo-Bede. *Libellus de astrolabio*. *PL* 90, 955–960. Paris, 1862.

Benedict of Nursia. *Regula Benedicti*. Ed. and trans. T. Fry. Collegeville, Minn., 1980.

Bernold of Constance. "Der Kalendar des Chronisten Bernold." Ed. R. Kuithan and J. Wollasch. *Deutsches Archiv für Erforschung des Mittelalters*, 40(1984):478–531.

Boethius. *De consolatione philosophiae*. Ed. H. F. Stewart. Loeb. Cambridge, Mass., 1936. *The Consolation of Philosophy*. Trans. R. Green. Indianapolis, 1962.

Byrhtferth of Ramsey. *Byrhtferth's Manual*. Ed. S. J. Crawford. EETS, 177. London, 1929.

Caesarius of Arles. *Sermones*. Ed. G. Morin. *CCSL* 103–104. Turnholt, 1953.

Calendar of Córdoba. *Le calendrier de Cordoue, Publié par R. Dozy, Nouvelle Édition*. Ed. C. Pellat. Leiden, 1961.

Calendar of Córdoba. "Una nueva traducción latina del Calendario de Córdoba (siglo XIII)." Ed. J. Martínez Gázquez and J. Samsó. Pp. 9–78 in J. Vernet, *Textos y estudios sobre astronomía Española*. Barcelona, 1981.

Campanus of Novara. *Theorica planetarum*. Ed. and trans. F. S. Benjamin, Jr., and G. J. Toomer. Madison, 1971.

Cassian, John. *De institutis coenobiorum*. Ed. J.-C. Guy. Sources Chrétiennes, 109. Paris, 1965.

Cassian, John. *Regula Cassiani*. In H. Ledoyen, "La 'Regula Cassiani' du CLM 28118 et la Règle anonyme de l'Escoriale A.I.13." *Revue Bénédictine*, 104(1984):154–194.

Cassiodorus Senator. *Expositio Psalmorum*. Ed. M. Adriaen. *CCSL* 97. Turnholt, 1958.

Cassiodorus Senator. *Institutiones*. Ed. R. A. B. Mynors. Oxford, 1937. *Introduction to the Divine and Human Readings*. Trans. L. W. Jones. Columbia University, Records of Civilization, 40. New York, 1940.

Cassiodorus Senator. *Variae*. Ed. T. Mommsen. *MGH*, Auct. Antiq. 12. Berlin, 1894.

*Catalogi bibliothecarum antiqui*. Ed. G. Becker. Bonn, 1885.

"Le Catalogue des manuscrits de l'Abbaye de Gorze au XI⁰ siècle." Ed. G. Morin. *Revue Bénédictine*, 22(1905):1–14.

Cetus Faventinus. *De diversis fabricis architectonicae.* Ed. V. Rose. Leipzig, 1899.

Chalcidius. [*In Timaeum.*] Ed. J. H. Waszink. *Corpus Platonicum Medii Aevi, Plato Latinus,* 4. London and Leiden, 1962.

Chrétien de Troyes. *Erec et Enide.* Trans. D. Gilbert. Berkeley, 1962.

Christian of Stavelot. *Expositio in Matthaeum. PL* 106, 1264–1594. Paris, 1864.

Clement of Alexandria. *Excerpta ex Theodoto.* Ed. F. Sagnard. Sources Chrétiennes, 23. Paris, 1948.

Cogitosus. *Sanctae Brigidae virginis vita. AASS,* Feb. I, 135–141. Paris, 1863. The edition in *PL* 72, 775–790 (Paris, 1864), omits several important passages. A new edition is in preparation by S. Connolly and J.-M. Picard; see their preliminary English translation, "Cogitosus: *Life of Saint Brigit,*" *Journal of the Royal Society of Antiquaries of Ireland,* 117(1987):11–27.

Columbanus. *Sancti Columbani opera.* Ed. and trans. G. S. M. Walker. Dublin, 1957.

*Commentariorum in Aratum reliquae.* Ed. E. Maas. Berlin, 1898.

*Concilia aevi Karolini.* Ed. A. Werminghoft. *MGH,* Conc., 2, 1. Hannover and Leipzig, 1906.

*Concilia Galliae, A. 314–A. 506, A. 511–A. 695.* Ed. C. de Clerq. *CCSL* 148, 148 A. Turnholt, 1963.

*Consuetudines saeculi X/XI/XII; Monumenta non-Cluniacensia, 1. Consuetudines Floriacenses antiquiores (saec. X).* Ed. A. Davril and L. Donnat. *CCM* 7, 3. Sieburg, 1984.

*Corpus Inscriptionum et Monumentorum Religionis Mithriacae,* 2 vols. Ed. M. J. Vermaseren. The Hague, 1956.

*Corpus Inscriptionum Latinarum,* Berlin, 1862– .

"Une correspondance d'écolâtres du onzième siècle." Ed. P. Tannery and Abbé Clerval. Pp. 229–303 in P. Tannery, *Mémoires Scientifiques,* 5, *Sciences exactes au Moyen Age.* Toulouse/Paris, 1922.

Cummian. *De controversia paschali.* Ed. and trans. M. Walsh and D. Ó Cróinín. Pontifical Institute of Medieval Studies, Studies and Texts, 86. Toronto, 1988.

pseudo-Cyprian. *De pascha computus.* Ed. G. Hartel. *CSEL* 3. 3. Vienna, 1871.

*De astrolabii compositione.* Pp. 308–315 in J. M. Millás Vallicrosa, *Assaig.* Barcelona, 1931.

*De mensura astrolabii.* Pp. 296–302 in J. M. Millás Vallicrosa, *Assaig.* Barcelona, 1931.

*De mensura astrolapsus.* Pp. 293–295 in J. M. Millás Vallicrosa, *Assaig.* Barcelona, 1931.

*De ratione conputandi.* Ed. D. Ó Cróinín. Pontifical Institute of Medieval Studies, Studies and Texts, 86. Toronto, 1988.

*De solsticia et aequinoctia conceptionis et nativitatis domini nostri Iesu Christi et Iohannis Baptistae.* In B. Botte, *Origines de la Noël.* Louvain, 1932.

*De utilitatibus astrolabii.* In Gerbert, *Opera Mathematica.* Hildesheim, 1963.

Dionysius Exiguus. *Praefatio ad Petronio De ratione pascha. PL* 67, 483–494. Paris, 1865.

*Divisio scientiarum.* In C. Lafleur, *Quatre introductions.* Montreal, 1988.

Dungal. *Epistolae.* Ed. E. Duemmler. *MGH,* Ep. 4, 563–585. Berlin, 1894.

Einhard. *Vita Karoli Magni.* Trans. E. S. Firchow and E. H. Zeydel. Coral Gables, Fla., 1972.

*English Kalendars before A.D. 1100.* Ed. Francis Wormald. Henry Bradshaw Society, 72. London, 1934.

Eriugena, Johannes Scottus. *Annotationes in Marcianum.* Ed. C. E. Lutz. Cambridge, Mass., 1939.

Eriugena, Johannes Scottus. *Periphyseon (De divisione naturae).* Ed. I. P. Sheldon-Williams. Scriptores Latini Hiberniae, 11. Dublin, 1981.

al-Farghānī, Abu'l-ʿAbbās Ahmad ibn Muhammad ibn Kathīr. *Rudimenta astronomica* in *Brevis ac perutilis compilatio Alfragani peritissimi*. . . . Nuremberg, 1537.

*Félire Oengusso: The Martyrology of Oengus the Culdee*. Ed. W. Stokes. Henry Bradshaw Society, 29. London, 1905.

Fulbert of Chartres. *The Letters and Poems of Fulbert of Chartres*. Ed. Frederick Behrends. Oxford, 1976.

Gerbert (Pope Sylvester II). *Constantino suo Gerbertus scolasticus*. In Gerbert, *Opera Mathematica*. Hildesheim, 1963.

Gerbert (Pope Sylvester II). *Gerberti postea Silvestri II papae Opera mathematica*. Ed. N. Bubnov. Hildesheim, 1963; repr. of the Berlin, 1899, ed.

Gerbert (Pope Sylvester II). *The Letters of Gerbert, with his Papal privileges as Sylvester II*. Trans. H. P. Lattin. Columbia University, Records of Civilization, 55. New York, 1961.

Germanicus Caesar. *Aratea cum scholiis*. Ed. A. Breysig. Hildesheim, 1967; repr. of the Berlin, 1867, ed.

*Glossa in Psalmos: The Hiberno-Latin Gloss on the Psalms of Codex Palatinus Latinus 68*. Ed. M. McNamara. Studi e Testi, 310. Vatican City, 1986.

Gregory of Tours. *De cursu stellarum ratio*. Ed. W. Arndt and B. Krusch. *MGH*, Scr. Rer. Merov. 1, 2. Hannover, 1885. Also Ed. F. Haase. Wroclaw, 1853.

Gregory of Tours. *Historia Francorum*. Trans. O. M. Dalton. Oxford, 1927.

Gregory the Great. *Homiliarum in evangelia*. PL 76, 1075–1312. Paris, 1878.

Gregory the Great. *Moralia in Iob*. Ed. M. Adriaen. CCSL 143. Turnholt, 1979–85.

Grosseteste, Robert. *Compotus . . . factus ad correctionum communis kalendarii nostri*. Ed. R. Steele. *Opera hactenus inedita Rogeri Baconi*, Fasc. 6. Oxford, 1926.

Grosseteste, Robert. *De sphaera*. Ed. L. Baur. *Die philosophischen werke*. Münster i. W., 1912.

Grosseteste, Robert. "Die Neumondtafel des Robertus Lincolniensis." Ed. A. Lindhagen. *Arkiv för matematik, astronomi och fysik*, 11,2(1916–17):1–41.

Grosseteste, Robert. *Die philosophischen Werke des Robert Grosseteste, Bischofs von Lincoln*. Ed. L. Baur. Münster i. W., 1912.

Helperic of Auxerre *Liber de computo*. PL 137, 17–48. Paris, 1879.

Hermann of Reichenau. *De mensura astrolabii*. J. Drecker, "Hermannus Contractus über das Astrolab." *Isis* 16(1931):200–219

Hesiod. *Works and Days*. Trans. A. N. Athanassakis. Baltimore, 1983.

Hildemar, *Expositio Regulae*. Ed. R. Mittermüller. Regensburg, New York, and Cincinnati, 1880.

*Horologium stellare monasticum* (saec. XI). *Consuetudines Benedictinae variae (saec. XI–saec. XIV)*. Ed. G. Constable. CCM 6, 16–18. Sieburg, 1975.

Hrabanus Maurus. *De clericorum institutione*. PL 107, 293–420. Paris, 1864.

Hrabanus Maurus. *De computo*. Ed. W. Stevens. CCCM 44. Turnholt, 1989.

Hrabanus Maurus. *De consanguineorum nuptiis et de magorum præstigiis falsisque divinationibus tractatus*. PL 110, 1087–1096. Paris, 1864.

Hrabanus Maurus. *De magicis artibus*. PL 110, 1095–1110. Paris, 1864.

Hrabanus Maurus. *Enarratio super Deuteronomium*. PL 108, 837–998. Paris, 1864.

Hrabanus Maurus. *Homiliae*. PL 110, 9–468. Paris, 1864.

Hrabanus Maurus. *In Genesim*. PL 107, 439–666. Paris, 1864.

Hrabanus Maurus. *In Matthaeum*. PL 107, 727–1156. Paris, 1864.

Hrabanus Maurus. *Martyrologium*. Ed. J. McCulloh. CCCM 44. Turnholt, 1989.

Hrabanus Maurus. *Poenitentium liber ad Otgarium*. PL 112, 1397–1430. Paris, 1878.

Ibn al-Muthannā, Muhammed ibn ʿAbd al-Karīm. *El comentario de Ibn al-Mutannā' a las Tabulas Astronómicas de al-Jwārizmī*. Ed. E. Millás Vendrell. Madrid and Barcelona, 1963.

Ibn al-Muthannā, Muhammed ibn ʿAbd al-Karīm. *Ibn al-Muthannâ's Commentary on the Astronomical Tables of al-Khwârizmî: Two Hebrew Versions, Edited and Translated, with an Astronomical Commentary*. Ed. B. R. Goldstein. New Haven, 1967.

Isidore of Seville. *De natura rerum*. Ed. J. Fontaine, *Traité de la Nature*. Bordeaux, 1960.

Isidore of Seville. *Etymologiae sive Originum*. Ed. W. M. Lindsay. Oxford, 1911.

Isidore of Seville. "The 'Institutionum Disciplinae' of Isidore of Seville." Ed. P. Pascal. *Traditio*, 13(1957):425–431.

Isidore of Seville. *Regula monachorum*. PL 83, 867–894. Paris, 1862.

Jerome. *Commentariorum in Amos*. Ed. M. Adriaen. *CCSL* 76. Turnholt, 1969.

Jerome. *Commentariorum in Esaiam*. Ed. M. Adriaen. *CCSL* 73. Turnholt, 1963.

Jerome. *Commentariorum in Hiezechielem*. Ed. F. Glorie. *CCSL* 75. Turnholt, 1964.

Jerome. *Commentariorum in Matheum*. Ed. D. Hurst and M. Adriaen. *CCSL* 77. Turnholt, 1969.

Jerome. *In die dominica paschae*. Ed. G. Morin. *Anecdota Maredsolana*, vol. 3, pars. 2. Maredsous/Oxford, 1897.

Jerome. *In Hieremiam libri VI*. Ed. S. Reiter. *CCSL* 74. Turnholt, 1960.

al-Khwārizmī, Muhammad ibn Mūsā. *Die astronomischen Tafeln des Muhammed ibn Mūsā al-Khwārizmī in der . . . latein. Uebersetzung des Athelard von Bath. . . .* Ed. H. Suter, with A. Bjørnbo and R. Besthorn. Kgl. Danske Vidensk. Selsk. Skrifter, 7. Række, Historisk og filosofisk Afd. 3, 1. Copenhagen, 1914. *The Astronomical Tables of al-Khwārizmī, Translation with Commentaries. . . .* Trans. O. Neugebauer. Det Koneglige Danske Videnskabernes Selskab, Hist. Filos. Skr., 4, 2. Copenhagen, 1962.

Lanfranc of Canterbury. *Monastic Constitutions*. Ed. and trans. D. Knowles. New York, 1951.

Lupitus of Barcelona(?) *Fragmentum libelli de astrolabio*. pp. 271–275. in J. M. Millás Vallicrosa, *Assaig*. Also pp. 370–375 in Gerbert, *Opera Mathematica*. Hildesheim, 1963.

Macrobius. *Commentarii in Somnium Scipionis*. Ed. J. Willis. Leipzig, 1970. *Macrobius' Commentary on the Dream of Scipio*. Trans. W. Stahl. New York, 1952.

Macrobius. *Saturnalia*. Trans. P. J. Davies. Columbia University, Records of Civilization, 79. New York, 1967.

Martianus Capella. *De nuptiis Philologiae et Mercurii*. Ed. J. Willis. Leipzig, 1983. *Martianus Capella and the Seven Liberal Arts*, vol. 2, *The Marriage of Philology and Mercury*. Trans. W. Stahl. Columbia University, Records of Civilization, 84. New York, 1977.

Martin of Braga. *Canones ex orientalium patrum synodis*. In *Opera omnia*. New Haven, 1950.

Martin of Braga. *De correctione rusticorum*. In *Opera omnia*. New Haven, 1950.

Martin of Braga. *De pascha*. In *Opera omnia*. New Haven, 1950.

Martin of Braga. *Martini episcopi Bracarensis Opera omnia*. Ed. C. W. Barlow. New Haven, 1950.

*Le Martyrologe d'Adon; ses deux familles, ses trois recensions*. Ed. J. Dubois and G. Renaud. Paris, 1984.

*Martyrology of Tallaght*. Ed. R. I. Best and H. J. Lawlor. Henry Bradshaw Society, 48. London, 1931.

Maximus of Turin. *Sermones*. Ed. A. Mutzenbecker. *CCSL* 23. Turnholt, 1962.

pseudo-Messahalla. *De compositione astrolabii; de operatione vel utilitate astrolabii*. In R. T. Gunther, *Early Science in Oxford*. Vol. 5. Oxford, 1929.

*Metrical Dindshenchas.* 5 vols. Ed. E. Gwynn. Dublin, 1991.

Michael Scot. *Commentary on The Sphere.* pp. 255–284 in L. Thorndike, *The* Sphere *of Sacrobosco.* Chicago, 1949.

"Milan Glosses on the Psalms." Ed. W. Stokes and J. Strachan. *Thesaurus Palaeohibernicus,* 1, *Biblical Glosses and Scholia.* Cambridge, 1901.

*Medieval Mensuration:* Quadrans Vetus *and* Geometrie Due Sunt Partes Principales. . . . Ed. N. L. Hahn. *Transactions of the American Philosophical Society,* vol. 72, pt. 8. Philadelphia, 1982.

Notker Balbulus of St. Gall. *De interpretibus divinarum scripturarum liber. PL* 131, 992–1004. Paris, 1884.

Pascasius Radbertus. *Expositio in Matheo libri XII.* Ed. B. Paul. *CCCM* 56. Turnholt, 1984.

*Pauli Warnefridi diaconi Casinensis in Sanctam Regulam Commentarium.* Typis Abbatiae Montis Casini, 1880.

Paulus Orosius. *Commonitorum de errore Priscillianistarum.* . . . Ed. G. de Christi, G. De Hartel, and G. Wattenbach. *CSEL* 18. Vienna, Prague, and Leipzig, 1889.

Pecham, John. *Tractatus de sphera.* Ed. R. B. MacLaren. Ph.D. diss. University of Wisconsin – Madison, 1978.

Peter Damian. *De perfectione monachorum. PL* 145, 291–328. Paris, 1867.

Peter Nightingale. *Petri Philomenae de Daciae et Petri de S. Audomaro, Opera quadrivialia.* Ed. F. S. Pedersen. Corpus philosophorum Danicorum medii aevi, 10.1–2. Copenhagen, 1983–84.

Petrus Alfonsi. *The* Disciplina Clericalis *of Petrus Alfonsi.* Trans. E. Hermes. Berkeley, 1977.

Petrus Alfonsi. *Epistola.* In J. M. Millás-Vallicrosa, "La aportación Astronómica de Pedro Alfonso." *Sefarad (Revista de Estudios Hebraicos),* 3(1943):65–105.

*Philosophia disciplina* (ca. 1245). Ed. C. Lafleur. *Quatre introductions.*

Pirmin of Reichenau. *De singulis libris canonicis scarapsus.* In G. Jecker, *Die Heimat des Hl. Pirmin, des Apostels der Alamannen.* Beiträge zur Geschichte des alten Mönchtums und des Benediktinerordens, 13. Münster i. W., 1927.

Plato. *Timaeus.*

Pliny. *Historia naturalis.* Trans. H. Rackham. Loeb. Cambridge, Mass., 1949.

Pomponius Mela. *De chorographia.* Ed. G. Ranstrand. Studia Graeca et Latina Gothoburgensia, 28. Stockholm, 1971.

*Priscillianist Tracts.* Ed. G. de Christi, G. De Hartel, and G. Wattenbach. *CSEL* 18. Vienna, Prague, and Leipzig, 1889.

*Prologus Cyrilli.* In B. Krusch, *84jährige Ostercyclus.* Leipzig, 1880.

Prudentius. *Apotheosis.* Trans. H. J. Thomson. Loeb. Cambridge, Mass., 1949.

Prudentius. *Cathemerinon.* Trans. H. J. Thomson. Loeb. Cambridge, Mass., 1949.

Prudentius. *Contra Orationem Symmachi.* Trans. H. J. Thomson. Loeb. Cambridge, Mass., 1953.

Ptolemy. *Almagest.* Trans. G. J. Toomer. New York, 1984.

Ptolemy. *The Arabic Version of Ptolemy's Planetary Hypothesis.* Ed. and trans. B. R. Goldstein. *Transactions of the American Philosophical Society,* vol. 57, pt. 4. Philadelphia, 1967.

Ptolemy. *Tables manuelles astronomiques de Ptolemée et de Théon.* 3 vols. (titles vary). Ed. N. B. Halma. Paris, 1822–25.

*Quatre introductions à la philosophie au XIII<sup>e</sup> Siècle.* Ed. C. Lafleur. Montreal, 1988.

Rabanus Maurus. *See* Hrabanus Maurus.

Raymond of Marseille. "Le Traité d'astrolabe de Raymond de Marseille." Ed. E. Poulle. *Studi medievali*, 3° ser., 5 (1964):866–905.

Reginald of Coldingham. *Vita S. Oswaldi regis et martyris*. In *Symeonis monachi Opera omnia*, vol. 1. Rolls Series, 75. Repr. Wiesbaden, 1965.

*Regula Magistri*. Ed. A. de Vogüé. Sources Chrétiennes, 105–106. Paris, 1964. *The Rule of the Master*. Trans. L. Eberle. Kalamazoo, Mich., 1977.

*Regularis concordia Anglicae nationis monachorum sanctimonialumque*. Ed. and trans. T. Symons. New York, 1933.

Remigius of Auxerre. In Boethius *De consolatione philosophiae*. There are three partial editions: "A Commentary by Remigius Autissiodorensis on the *De consolatione philosophiae* of Boethius," ed. H. F. Stewart, *The Journal of Theological Studies*, 17 (1915):22–42. "Le commentaire inédit de Jean Scot Érigène au Mètre IX du Livre III du 'De consolatione philosophiae' de Boèce," ed. H. Silvestre, *Revue d'Histoire Ecclésiastique*, 47(1953):44–122 (text at pp. 51–65). *Saeculi noni auctoris in Boetii Consolationem Philosophiae Commentarius*, ed. E. T. Silk, Papers and Monographs of the American Academy in Rome, 9 (Rome, 1935).

Remigius of Auxerre. *Commentum in Martianum Capellam*. 2 vols. Ed. C. E. Lutz. Leiden, 1962–65.

Remigius of Auxerre. *Enarrationes in Psalmos*. PL 131, 149–844. Paris, 1884.

Robertus Anglicus. *Commentary on The Sphere*. pp. 143–157 in L. Thorndike, *The Sphere of Sacrobosco*, Chicago, 1949.

Sacrobosco, John. *The Sphere of Sacrobosco and Its Commentators*. Ed. L. Thorndike. Chicago, 1949.

*Das Sakramentar-Pontifikale, des Bischofs Wolfgang von Regensburg* (Verona, Bibl. Cap., Cod. LXXXVII). Ed. K. Gamber and S. Rehle. Textus Patristici et Liturgici, 15. Regensburg, 1985.

*Sanas Chormaic, Cormac's Glossary*. Ed. W. Stokes, Trans. J. O'Donovan. Calcutta, 1868.

Seneca. *Naturales quaestiones*. Trans. H. Corcoran. Loeb. Cambridge, Mass., 1971–72.

Sidonius Apollinaris. *Epistulae*. Trans. W. B. Anderson. Loeb. Cambridge, Mass., 1936–65.

Sisebut. *Epistula ad Isidorum*. Ed. J. Fontaine. In *Isidore de Seville Traité de la Nature*. Bordeaux, 1960.

Smaragdus of St. Mihiel. *Smaragdi abbatis Exposito in Regulam S. Benedicti*. Ed. A. Spannagel and P. Englebert. *CCM* 8. Sieburg, 1974.

Theodore of Mopsuestia. *Expositionis in Psalmos*. Ed. L. de Coninck and M. J. d'Hont. *CCSL* 88 A. Turnholt, 1987.

Theon of Alexandria. *Le "petit commentaire" de Théon d'Alexandrie aux tables faciles de Ptolémée*. Ed. Anne Tihon. Studi e Testi, 282. Vatican City, 1978.

Theophilus of Alexandria. *Epistola ad Theodosium Augustum*. In B. Krusch, *84jährige Ostercyclus*. Leipzig, 1880.

*Ultan's Hymn*. Ed. W. Stokes and J. Strachan. *Thesaurus Palaeohibernicus*, 2. Cambridge, 1903.

Victorius of Aquitaine. *Prologus Victorii Aquitani ad Hilarum archdiaconum*. In B. Krusch, *Entstehung unserer heutigen Zeitrechnung*. Berlin, 1938.

Virgil. *Georgics*. Trans. H. R. Fairclough. Loeb. Cambridge, Mass., 1938.

*Vita Iohannis Gorziensis*. Ed. G. H. Pertz. *MGH*, SS 4, 335–377. Hannover, 1841. Repr. in PL 137, 239–310. Paris, 1879.

*Vita prolixior S. Justi episcopi Lugdunensis*. *AASS*, Sept. I, 373–374. Paris, 1868.

Vitruvius. *De architectura*. Trans. F. Granger. Loeb. Cambridge, Mass., 1931–34.

William of St. Cloud. *Kalendarium Regine*. Ed. R. Harper. Ph.D. diss., Emory University, 1966.

Zeno of Verona. *Tractatus*. Ed. B. Löfstedt. *CCSL* 22. Turnholt, 1971.


## Studies

Allen, Richard H. *Star Names: Their Lore and Meaning*. New York: Dover, 1963.

*Arts Libéraux et Philosophie au Moyen Âge*. Actes du quatrième Congrès international du philosophie médiévale. Montreal: Institut d'Études Médievales, 1969.

Ashtor, Eliyahu. *The Jews of Moslem Spain*. Vol. 1. Philadelphia: Jewish Publication Society of America, 1973.

Atkinson, R. J. C. "Aspects of the Archaeoastronomy of Stonehenge." Pp. 107–116 in Heggie, *Archaeoastronomy in the Old World*.

Audin, Amable. "L'Omphalos de Lugdunum." Pp. 152–164 in Marcel Renard, ed., *Hommages à Albert Grenier*, Collection Latomus, 58. Brussels: Latomus, 1962.

Aveni, Anthony F., ed. *Archaeoastronomy in the New World: American Primitive Astronomy*. Cambridge: Cambridge Univ. Pr., 1982.

Aveni, Anthony F., ed. *World Archaeoastronomy*. Cambridge: Cambridge Univ. Pr., 1989.

Baumstark, Anton. *Comparative Liturgy*. Westminster, Md.: Newman Press, 1958.

Baur, Ludwig. *Die Philosophie des Robert Grosseteste, Bischofs von Lincoln*. BGPMA, 18,4–6. Münster i. W.: Verlag Aschendorff, 1917.

Beaujouan, Guy. "The Transformation of the Quadrivium." Pp. 463–487 in Benson, Constable, and Latham, *Renaissance and Renewal*.

Beck, Roger. *Planetary Gods and Planetary Orders in the Mysteries of Mithras*. Études préliminaires aux religions orientales dans l'Empire romain. Leiden: E. J. Brill, 1988.

Benjamin, Francis, Jr., and G. J. Toomer, eds. and trans. *Campanus of Novara and Medieval Planetary Theory:* Theorica Planetarum. Madison: Univ. of Wisconsin Pr., 1971.

Benson, Robert L., Giles Constable, and Carol D. Latham. *Renaissance and Renewal in the Twelfth Century*. Cambridge: Harvard Univ. Pr., 1982.

Bergmann, Werner. *Innovationen im Quadrivium des 10. und 11. Jahrhunderts: Studien zur Einführung von Astrolab und Abakus im lateinischen Mittelalter, Sudhoffs Archiv, Zeitschrift für Wissenshaftsgeschichte, Beihefte*, 26. Stuttgart: Franz Steiner Verlag, 1985.

Bergmann, Werner. "Der römische Kalendar: Zur sozialen Konstruktion der Zeitrechnung: Ein Beitrag zur Soziologie der Zeit." *Saeculum*, 35(1984):1–16.

Bergmann, Werner, and Wolfhard Schlosser. "Gregor von Tours und der 'rote Sirius'; Untersuchungen zu den astronomischen Angaben in *De cursu stellarum ratio*." *Francia*, 15(1987):43–74.

*Bibliotheca Hagiographica Latina*. Brussels: Socii Bollandiani, 1898–99.

Bischoff, Bernard, Bruce Eastwood, Thomas A.-P. Klein, Florentine Mütherich, and Pieter F. J. Obbema. *Aratea: Kommentar zum Aratus des Germanicus, MS. Voss. Lat. Q. 79, Bibliotheek der Rijksuniversiteit Leiden*. Lucerne: Faksimile Verlag, 1989.

Bishop, Edmund. "On the Origins of the Feast of the Conception of the Blessed Virgin Mary." Pp. 238–259 in his *Liturgica Historica: Papers on the Liturgy and Religious Life of the Western Church*. Oxford: Clarendon Press, 1918.

Bitel, Lisa M. *Isle of the Saints: Monastic Settlement and Christian Community in Early Ireland.* Ithaca: Cornell Univ. Pr., 1990.

Bloch, Herbert. "The Pagan Revival in the West at the End of the Fourth Century." Pp. 193–218 in Momigliano, *Conflict Between Paganism and Christianity.*

Bonnaud, R. "Notes sur l'Astrologie Latine au VI$^e$ Siècle." *Revue Belge de Philologie et d'Histoire,* 10(1931):554–577.

Borst, Arno. "Alkuin und die Enzyklopädie von 809." Pp. 53–78 in Butzer and Lohrmann, *Science in Carolingian Times.*

Borst, Arno. *Astrolab und Klosterreform an der Jahrtausendwende, Sitzungsberichte der Heidelberger Akademie der Wissenschaften.* Philosophisch-historische Klasse, Jahrg. 1989, Bericht 1. Heidelberg: Universitätsverlag Carl Winter, 1989.

Borst, Arno. *Das Buch der Naturgeschichte: Plinius und seine Leser im Zeitalter des Pergaments, Abhandlungen der Heidelberger Akademie der Wissenschaften.* Philosophisch-historische Klasse, Abh. 2. Heidelberg: Universitätsverlag Carl Winter, 1994.

Borst, Arno. *The Ordering of Time: From the Ancient Computus to the Modern Computer.* Translated by Andrew Winnard. Cambridge: Polity Press; Chicago: Univ. of Chicago Pr., 1993.

Botte, Bernard. *Les origines de la Noël et de l'Épiphanie: Étude historique.* Textes et Études Liturgiques, 1. Louvain: Abbaye du Mont César, 1932.

Bowen, Emrys George. "The Cult of St. Brigit." *Studia Celtica,* 8/9(1973/74):33–47.

Brecher, Kenneth, and Michael Feirtag. *Astronomy of the Ancients.* Cambridge: MIT Press, 1979.

Bretagnon, Pierre, and Jean-Louis Simon. *Tables for the Motion of the Sun and the Five Bright Planets from −4000 to +2800.* Richmond, Va.: Willmann-Bell, 1986.

Broda, Johanna. "Significant Dates of the Mesoamerican Agricultural Calendar and Archaeoastronomy." Abstract, p. 494 in Aveni, *World Archaeoastronomy.*

Brown, Peter. *The Cult of the Saints: Its Rise and Function in Latin Christianity.* Chicago: Univ. of Chicago Pr., 1981.

Burl, Aubrey. " 'By the Light of the Cinerary Moon': Chambered Tombs and the Astronomy of Death." Pp. 243–274 in Ruggles and Whittle, *Astronomy and Society in Britain.*

Burl, Aubrey. " 'Without Sharp North . . . ': Alexander Thom and the Great Stone Circles of Cumbria." Pp. 175–205 in Ruggles, *Records in Stone.*

Burnett, Charles, ed. *Adelard of Bath: An English Scientist and Arabist of the Early Twelfth Century.* Warburg Institute Surveys and Texts, 14. London: University of London, 1987.

Butzer, Paul L., and Dietrich Lohrmann. *Science in Western and Eastern Civilization in Carolingian Times.* Basel, Boston, and Berlin: Birkhäuser Verlag, 1993.

Carmody, Francis J. *Arabic Astronomical and Astrological Sciences in Latin Translation: A Critical Bibliography.* Berkeley: Univ. of California Pr., 1956.

Chadwick, Henry. *Priscillian of Avila: The Occult and the Charismatic in the Early Church.* Oxford: Clarendon Press, 1976.

Chadwick, Nora K. *The Druids.* Cardiff: Univ. of Wales Pr., 1966.

Chadwick, Owen. *John Cassian.* Cambridge: Cambridge Univ. Pr., 1968.

Chaney, William A. *The Cult of Kingship in Anglo-Saxon England: The Transition from Paganism to Christianity.* Berkeley: Univ. of California Pr., 1970.

Coebergh, C. "L'Épiphanie à Rome Avant Saint Léon: Un indice pour l'année 419." *Revue Bénédictine*, 75(1965):304–307.

Coyne, G. V., M. A. Hoskin, and O. Pedersen, eds. *Gregorian Reform of the Calendar*. Proceedings of the Vatican Conference to commemorate its 400th anniversary. Vatican: Specolo Vaticana, 1983.

Cumont, Franz. *The Mysteries of Mithra*. Translated by Thomas J. McCormack. New York: Dover, 1956.

d'Alverny, Marie-Thérèse. "Translations and Translators." Pp. 421–462 in Benson, Constable, and Latham, *Renaissance and Renewal*.

d'Haenens, Albert. "La clepsydre de Villers (1267)." Pp. 321–342 in *Klösterliche Sachkultur des Spätmittelalters, Sitzungsberichte der österreichische Akademien der Wissenschaften*. Philosophisch-historische Klasse, 367. Vienna, 1980.

Dales, Richard C. "The Computistical Works Ascribed to Robert Grosseteste." *Isis*, 80(1989):74–79.

Daniélou, Jean. "Les Douze Apôtres et le Zodiaque." *Vigiliae Christianae*, 13(1959):14–21.

Daniélou, Jean. *The Theology of Jewish Christianity*. Translated and edited by John A. Baker. London: Darton Longman & Todd, 1964.

Daviet, Roger. "La Mesure du Temps en Gaule: Essai d'Interprétation du Calendrier de Coligny." *Revue Archéologique de l'Est et du Centre-Est*, 14(1963):53–80.

de Nie, Giselle. "Roses in January: A Neglected Dimension in Gregory of Tours' *Historiae*." *Journal of Medieval History*, 5(1979):259–289.

de Nie, Giselle. "The Spring, the Seed, and the Tree: Gregory of Tours on the Wonders of Nature." *Journal of Medieval History*, 11(1985):89–135.

de Nie, Giselle. *Views from a Many-windowed Tower: Studies of Imagination in the Works of Gregory of Tours*. Amsterdam: Rodopi, 1987.

de Vries, Jan. *Keltische Religion*. Die Religionen der Menschheit 18. Stuttgart: W. Kohlhammer Verlag, 1961.

Delahaye, Philippe. "La Place des Arts Libéraux dans les Programmes Scolaires du XIII$^e$ Siècle." Pp. 161–173 in *Arts Libéraux*.

Destombes, Marcel. "Un astrolabe carolingien et l'origine de nos chiffres arabes." *Archives internationales d'histoire des sciences*, 58/59:(1962):3–45

Destombes, Marcel. *Mappemondes, A.D. 1200–1500*. Monumenta Cartographica Vetustioris Aevi, 1. Amsterdam: N. Israel, 1964.

Díaz y Díaz, Manuel C. "Les Arts Libéraux d'Apres les Écrivains Espagnols et Insulaires aux VII$^e$ et VIII$^e$ Siècles." Pp. 37–46 in *Arts Libéraux*.

Dickey, Bruce G. "Adelard of Bath. An Examination Based on Heretofore Unexamined Manuscripts." Ph.D. diss., University of Toronto, 1983.

Dieterich, Albrecht. "Die Weisen aus dem Morgenlande." *Zeitschrift für die neutestamentliche Wissenschaft*, 3(1902): 1–14.

Dölger, Franz Joseph. "Die Planetenwoche der griechisch-römischen Antike und der christliche Sonntage." *Antike und Christentum*, 6(1940):202–238.

Dölger, Franz Joseph. *Sol Salutis: Gebet und Gesang im christlichen Altertum: mit besonderer Rucksicht auf die Ostung in Gebet und Liturgie*. Liturgiegeschichtliche Forschungen, 4/5. Münster i. W.: Verlag Aschendorff, 1920.

Duhem, Pierre. *Le Système du Monde*. 10 vols. Paris: Hermann, 1914–58.

Duval, Paul-Marie, and Georges Pinault. *Recueil des inscriptions gauloises*, RIG, vol. 3, les calendriers, Coligny, Villards d'Héiria. Paris: Éditions du Centre Nationale de la Recherche Scientifique, 1986.

Eastwood, Bruce S. "The Astronomies of Pliny, Martianus Capella, and Isidore of Seville in the Carolingian World." Pp. 169–174 in Butzer and Lohrmann, *Science in Carolingian Times.*

Eastwood, Bruce S. *Astronomy and Optics from Pliny to Descartes: Texts, Diagrams, and Conceptual Structures.* Variorum Collected Studies, CS 291. London: Variorum, 1989.

Eastwood, Bruce S. "The Astronomy of Macrobius in Carolingian Europe: Dungal's Letter of 811 to Charles the Great." *Early Medieval Europe,* 3(1994):117–134.

Eastwood, Bruce S. "Origins and Contents of the Leiden Planetary Configuration (ms Voss. Q.79, fol. 93v), an Artistic Astronomical Schema of the Early Middle Ages." *Viator,* 14(1983):1–40

Eastwood, Bruce S. "Plinian Astronomical Diagrams in the Early Middle Ages." Pp. 141–172 in Grant and Murdoch, *Mathematics and Its Applications.*

Eastwood, Bruce S. "Plinian Astronomy in the Middle Ages and the Renaissance." Pp. 197–235 in French and Greenaway, *Pliny, His Sources and Influence.*

Eliade, Mircea. *Cosmos and History: The Myth of the Eternal Return.* Translated by Willard R. Trask. New York: Harper & Row, 1959.

Eliade, Mircea. *The Sacred and the Profane: The Nature of Religion.* Translated by Willard R. Trask. New York: Harcourt, Brace, & World, 1959.

Eriksson, Sven. *Wochentagsgötter, Mond und Tierkreis: Laienastrologie in der römischen Kaiserzeit.* Studia Graeca et Latina Gothoburgensia, 3. Stockholm: Almqvist & Wiksell, 1956.

Estrey, F. N. "Charlemagne's Silver Celestial Table." *Speculum,* 18(1943):112–117.

Evans, Gillian R. "Schools and Scholars: The Study of the Abacus in English Schools c. 980–c. 1150." *English Historical Review,* 94(1979):71–89.

Evans, G. R., and A. M. Peden. "Natural Science and the Liberal Arts in Abbo of Fleury's Commentary on the Calculus of Victorius of Aquitaine." *Viator,* 16(1985):108–127.

*Explanatory Supplement to the Astronomical Ephemeris. . . .* London: HMSO, 1961.

Flint, Valerie I. J. *The Rise of Magic in Early Medieval Europe.* Princeton: Princeton Univ. Pr., 1991.

Folkerts, Menso. *"Boethius" Geometrie II: Ein mathematisches Lehrbuch des Mittelalters.* Wiesbaden: F. Steiner, 1970.

Folz, Robert. "Saint Oswald Roi de Northumbrie: Étude d'hagiographie royale." *Analecta Bollandiana,* 98(1980):49–74.

Folz, Robert. *Les saints rois du Moyen Age en Occident (VI<sup>e</sup>–XIII<sup>e</sup> siècles).* Subsidia Hagiographica, 68. Brussels: Société des Bollandistes, 1984.

Frank, Hieronymus. "Frühgeschichte und Ursprung des römischen Weihnachtsfestes im Lichte neuerer Forschung." *Archiv für Liturgiewissenschaft,* 2(1952):1–24.

French, Roger, and Frank Greenaway. *Science in the Early Roman Empire: Pliny the Elder, His Sources and Influence.* Totowa, N.J.: Barnes & Noble Books, 1986.

Geary, Patrick J. *Before France and Germany: The Creation and Transformation of the Merovingian World.* New York: Oxford Univ. Pr., 1988.

Gibbs, Sharon L. *Greek and Roman Sundials.* New Haven: Yale Univ. Pr., 1976.

Gibbs, Sharon L., and George Saliba. *Planispheric Astrolabes from the National Museum of American History.* Smithsonian Studies in History and Technology, 45. Washington: Smithsonian Institution Press, 1984.

Gibson, Margaret. "Adelard of Bath." Pp. 7–16 in Burnett, *Adelard of Bath*.

Gibson, Margaret, ed. *Boethius: His Life, Thought, and Influence*. Oxford: Basil Blackwell, 1981

Gibson, Margaret, and Janet Nelson, eds. *Charles the Bald: Court and Kingdom*. BAR International Series, 101. Oxford: B.A.R., 1981.

Gingerich, Owen. "The Basic Astronomy of Stonehenge." Pp. 117–132 in Brecher and Feirtag, *Astronomy of the Ancients*.

Goldstine, Herman H. *New and Full Moons, 1001 B.C. to A.D. 1651, Memoirs of the American Philosophical Society*, 94. Philadelphia, 1973.

Gougaud, Louis. *Les Saints irlandais hors d'Irlande: Études dans le culte et dans la dévotion traditionelle*. Bibliothèque de la Revue d'Histoire Ecclésiastique, 16. Louvain and Oxford, 1936.

Grant, Edward. *Planets, Stars, and Orbs: The Medieval Cosmos, 1200–1687*. Cambridge: Cambridge Univ. Pr., 1994.

Grant, Edward, and John E. Murdoch, eds. *Mathematics and Its Applications to Science and Natural Philosophy in the Middle Ages*. Cambridge: Cambridge Univ. Pr., 1987.

Grosjean, Paul. "Notes d'hagiographie celtique: 1. La prétendue fête de la Conception de la Sainte Vierge dans les églises celtiques." *Analecta Bollandiana*, 61(1943):91–95.

Grosjean, Paul. "Recherches sur les debuts de la controverse Pascale chez les Celtes." *Analecta Bollandiana*, 64(1946):200–243.

Harper, Richard I. "The *Kalendarium Regine* of Guillaume de St.-Cloud." Ph.D. diss., Emory University, 1966.

Harrison, Kenneth. "Episodes in the History of Easter Cycles in Ireland." Pp. 307–319 in Whitelock, McKitterick, and Dumville, *Ireland in Early Medieval Europe*.

Harrison, Kenneth. "Luni-Solar Cycles: Their Accuracy and Some Types of Usage." Pp. 65–78 in King and Stevens, *Saints, Scholars, and Heroes*, vol. 2.

Haskins, Charles Homer. *Studies in the History of Mediaeval Science*. New York: Frederick Ungar, 1967.

Hawkins, Gerald S., with John B. White. *Stonehenge Decoded*. Garden City, N.Y.: Doubleday & Co., 1965.

Heggie, Douglas C. *Megalithic Science: Ancient Mathematics and Astronomy in Northwest Europe*. London: Thames & Hudson, 1981.

Heggie, Douglas C., ed. *Archaeoastronomy in the Old World*. Cambridge: Cambridge Univ. Pr., 1982.

Hermes, Eberhard. *The Disciplina Clericalis of Petrus Alfonsi*. Berkeley: Univ. of California Pr., [1977].

Hollis, Martin, and Steven Lukes, eds. *Rationality and Relativism*. Cambridge: MIT Press, 1989.

Honigmann, Ernst. *Die sieben Klimata und die ΠΟΛΕΙΣ ΕΠΙΣΗΜΟΙ*. Heidelberg: Universitätsverlag Carl Winter, 1929.

Horton, Robin, and Ruth Finnegan, eds. *Modes of Thought: Essays on Thinking in Western and Non-Western Societies*. London: Faber & Faber, 1973.

Hübner, Wolfgang. "Das Horoskop der Christen: (Zeno 1,38 L.)." *Vigiliae Christianae*, 29(1975):120–137.

Hudson, Travis. "California's First Astronomers." Pp. 11–81 in Krupp, *Archaeoastronomy and the Roots of Science*.

Hughes, Kathleen. *The Church in Early Irish Society*. Ithaca: Cornell Univ. Pr., 1966.

Hughes, Kathleen. *Early Christian Ireland: Introduction to the Sources*. Ithaca: Cornell Univ. Pr., 1972.

Jones, Charles W. *Bedae Pseudepigrapha: Scientific Writings Falsely Attributed to Bede*. Ithaca: Cornell Univ. Pr., 1939.

Jones, Charles W. *Bede, the Schools, and the Computus*. Edited by Wesley M. Stevens. Variorum Collected Studies, CS 436. Aldershot: Variorum, 1994.

Jones, Charles W. "The 'Lost' Sirmond Manuscript of Bede's Computus." *English Historical Review*, 52(1937):204–219.

Jones, Charles W. "The Victorian and Dionysiac Paschal Tables in the West." *Speculum*, 9(1934):408–421.

Kantorowicz, E. H. "Oriens Augusti – Lever du Roi." *Dumbarton Oaks Papers*, 17(1963): 119–177.

Kantorowicz, E. H. "Puer Exoriens." Pp. 25–36 in his *Selected Studies*. Locust Valley, N.Y.: J. J. Augustin, 1965.

Kennedy, E. S. *A Survey of Islamic Astronomical Tables, Transactions of the American Philosophical Society*, 46, 2. Philadelphia, 1956.

Kenney, James F. *The Sources for the Early History of Ireland: An Introduction and Guide*. Vol. 1. New York: Columbia Univ. Pr., 1929.

Kibre, Pearl. "The *Quadrivium* in the Thirteenth Century Universities (with Special Reference to Paris)." Pp. 175–191 in *Arts Libéraux*.

King, David A. *Astronomy in the Service of Islam*. Variorum Collected Studies, CS 416. Aldershot: Variorum, 1993.

King, David A. "Folk Astronomy in the Service of Religion: The Case of Islam." Pp. 124–138 in Ruggles and Saunders, *Astronomies and Cultures*.

King, David A., and George Saliba, eds. *From Deferent to Equant: Studies in the History of Science in the Ancient and Medieval Near East in Honor of E. S. Kennedy, Annals of the New York Academy of Sciences*, 500. New York, 1987.

King, Margot H., and Wesley M. Stevens, eds. *Saints, Scholars, and Heroes: Studies in Medieval Culture in Honour of Charles W. Jones*. 2 vols. Collegeville, Minn.: Saint John's Abbey and University, 1979.

King, Vernon H. "An Investigation of Some Astronomical Excerpts from Pliny's Natural History Found in Manuscripts of the Earlier Middle Ages." B. Litt thesis, Oxford University, 1969.

Krupp, E. C., ed. *Archaeoastronomy and the Roots of Science*. AAAS Selected Symposium, 71. Boulder, Colo.: Westview Press, 1984.

Krusch, Bruno. *Studien zur christlich-mittelalterlichen Chronologie*, [1] *Der 84jährige Ostercyclus und seine Quellen*. Leipzig: von Zeit & comp., 1880.

Krusch, Bruno. *Studien zur christlich-mittelalterlichen Chronologie*, [2] *Die Entstehung unserer heutigen Zeitrechnung, Abhandlungen der Preußischen Akademie der Wissenschaften*, 8, 1937. Berlin, 1938.

Kunitzsch, Paul. *Der Almagest: Die Syntaxis Mathematica des Claudius Ptolemäus in arabisch-lateinischer Überlieferung*. Wiesbaden: Otto Harrasowitz, 1974.

Kunitzsch, Paul. *Typen von Sternverzeichnissen in astronomischen Handschriften des zehnten bis vierzehnten Jahrhunderts*. Wiesbaden: Otto Harrasowitz, 1966.

Kunitzsch, Paul. *Über eine anwā'-Tradition mit bisher unbekannten Sternnamen, Sitzungsberichte der Bayerische Akademie der Wissenschaften.* Philosophisch-historische Klasse, Jahrg. 1983, Heft 5. Munich, 1983.

Ladner, G. B. "Terms and Ideas of Renewal." Pp. 1–33 in Benson, Constable, and Latham, *Renaissance and Renewal.*

Lafleur, Claude. *Quatre introductions à la philosophie au XIIIᵉ Siècle.* Montreal: Institut d'Études Médiévales, 1988.

Laistner, M. L. W. "Antiochene Exegesis in Western Europe during the Middle Ages." *Harvard Theological Review,* 40(1947):29.

Laistner, M. L. W. *The Intellectual Heritage of the Early Middle Ages.* Ithaca: Cornell Univ. Pr., 1957.

Laistner, M. L. W. "Some Early Medieval Commentaries on the Old Testament." *Harvard Theological Review,* 46(1953):27–46, repr. in his *The Intellectual Heritage.*

Laistner, M. L. W. *Thought and Letters in Western Europe, A.D. 500–900.* Ithaca: Cornell Univ. Pr., 1957.

Laistner, M. L. W. "The Western Church and Astrology during the Early Middle Ages." *Harvard Theological Review,* 34(1941):251–275, repr. in his *The Intellectual Heritage.*

LeMay, Richard. *Abu MaʿShar and Latin Aristotelianism in the Twelfth Century: The Recovery of Aristotle's Natural Philosophy through Arabic Astrology,* American University of Beirut. Publication of the Faculty of Arts and Sciences, Oriental Series, 38. Beirut, 1962.

LeMay, Richard. "Dans l'Espagne du XIIᵉ Siècle: Les Traductions de l'Arabe au Latin." *Annales: Economies, Sociétés, Civilisations,* 18(1963):639–665.

Lindberg, David C. *Science in the Middle Ages.* Chicago: Univ. of Chicago Pr., 1978.

Lockyer, Joseph Norman. *Stonehenge and Other British Stone Monuments Astronomically Considered.* London: Macmillan, 1906.

Lowe, E. A., ed. *Codices Latini Antiquiores.* 12 vols. Oxford: Clarendon Press, 1934–66.

Lowrie, Walter. *Art in the Early Church.* New York: Pantheon Books, 1947.

MacCana, Proinsias. *Celtic Mythology.* New York: Peter Bedrick Books, 1983.

MacCana, Proinsias. "Celtic Religion." In M. Eliade, ed., *The Encyclopedia of Religion.* New York: Macmillan, 1987.

McCarthy, Daniel. "Easter Principles and a Fifth-century Lunar Cycle Used in the British Isles." *Journal for the History of Astronomy,* 24 (1993):204–224.

McCarthy, Daniel, and Dáibhí Ó Cróinín. "The 'Lost' Irish 84-year Easter Table Rediscovered." *Peritia,* 6/7 (1987/88):227–242.

McCluskey, Stephen. "Calendars and Symbolism: Functions of Observation in Hopi Astronomy." *Archaeoastronomy,* no. 15; suppl. to *Journal for the History of Astronomy,* 21(1990):S1– S16.

McCluskey, Stephen. "Gregory of Tours, Monastic Timekeeping, and Early Christian Attitudes to Astronomy." *Isis,* 81(1990):8–22.

McCluskey, Stephen. "Historical Archaeoastronomy: The Hopi Example." Pp. 31–57 in Aveni, *Archaeoastronomy in the New World.*

McCluskey, Stephen. "The Mid-Quarter Days and the Historical Survival of British Folk Astronomy." *Archaeoastronomy,* no. 13; suppl. to *Journal for the History of Astronomy,* 20(1989):S1–S19.

McCluskey, Stephen. "The Solar Year in the Calendar of Coligny." *Études Celtiques,* 27(1990):163–174.

McCone, Kim. "Brigit in the Seventh Century: A Saint with Three Lives?" *Peritia,* 1(1982): 107–145.

MacCormack, Sabine G. *Art and Ceremony in Late Antiquity*. Berkeley: Univ. of California Pr., 1981.

McCulloh, John. "*Martyrologium Excarpsatum*: A New Text from the Early Middle Ages." Pp. 179–237 in King and Stevens, *Saints, Scholars, and Heroes*, vol. 2.

McGurk, Patrick. "Carolingian Astronomical Manuscripts." Pp. 317–332 in Gibson and Nelson, *Charles the Bald*.

McGurk, Patrick. "Computus Helperici: Its Transmission in England in the Eleventh and Twelfth Centuries." *Medium Aevum*, 43(1974):1–5

McGurk, Patrick. "The Metrical Calendar of Hampson: A New Edition." *Analecta Bollandiana*, 104(1986):79–125.

MacKie, Euan. *Science and Society in Prehistoric Britain*. New York: St. Martin's Press, 1977.

McKinney, Loren C. *Bishop Fulbert and Education at the School of Chartres*. Texts and Studies in the History of Medieval Education, 6. Notre Dame, Ind.: The Medieval Institute, 1957.

McKitterick, Rosamond. *The Frankish Kingdoms under the Carolingians, 751–987*. New York: Longman, 1983.

McNally, Robert E. "The Three Holy Kings in Early Irish Writing." Vol. 2, pp. 667–690, in P. Granfield and J. A. Jungmann, eds., *Kyriakon: Festschrift Johannes Quasten*. Münster i. W.: Verlag Aschendorff, 1970.

MacNeill, Eóin. "On the Notation and Chronography of the Calendar of Coligny." *Ériu*, 10(1926):1–67.

MacNeill, Máire. *The Festival of Lughnasa: A Study of the Survival of the Celtic Festival of the Beginning of Harvest*. London: Oxford Univ. Pr., 1962.

Malotki, Ekkehart. *Hopi Time: A Linguistic Analysis of the Temporal Concepts in the Hopi Language*. Trends in Linguistics: Studies and Monographs. Berlin and New York: Mouton, 1983.

Mancha, J. L. "Astronomical Use of Pinhole Images in William of Saint-Cloud's *Almanach Planetarum* (1292)." *Archive for History of Exact Sciences*, 43(1992):275–298.

Mercier, Raymond. "Astronomical Tables in the Twelfth Century." Pp. 87–115 in Burnett, *Adelard of Bath*.

Millás Vallicrosa, Jose Maria. "La aportación Astronómica de Pedro Alfonso." *Sefarad (Revista de Estudios Hebraicos)*, 3(1943):65–105.

Millás Vallicrosa, Jose Maria. *Assaig d'Història de les idees fisiques i matemàtiques a la Catalunya Medieval*. Estudis Universitaris Catalans, 1. Barcelona, 1931.

Moesgaard, Kristian Peder. "The Full Moon Serpent: A Foundation Stone of Ancient Astronomy?" *Centaurus*, 24(1980):51–96.

Moir, Gordon. "Some Archaeological and Astronomical Objections to Scientific Astronomy in British Prehistory." Pp. 221–241 in Ruggles and Whittle, *Astronomy and Society in Britain*.

Momigliano, Arnaldo, ed. *The Conflict Between Paganism and Christianity in the Fourth Century*. Oxford: Clarendon Press, 1965.

Mostert, Richard, and Marco Mostert. "Using Astronomy as an Aid to dating Manuscripts: The Example of the Leiden Aratea Planetarium." *Quaerendo*, 20(1990):248–261.

Murschel, Andrea. "The Structure and Function of Ptolemy's Physical Hypotheses of Planetary Motion." *Journal for the History of Astronomy*, 26(1995):33–61.

Neugebauer, Otto. *Ethiopic Astronomy and Computus, Sitzungsberichte der österreichische Akademie der Wissenschaften*, 347. Vienna, 1979.

Neugebauer, Otto. *A History of Ancient Mathematical Astronomy*. 3 vols. Berlin, Heidelberg, and New York: Springer-Verlag, 1975.

Neugebauer, Otto. "The Transmission of Planetary Theories in Ancient and Medieval Astronomy." *Scripta Mathematica*, 22(1956):165–192.

Nilsson, Martin P. *Primitive Time-Reckoning: A Study in the Origins and First Development of the Art of Counting Time*. . . . Lund: C. W. K. Gleerup, 1920.

Noble, Joseph, and Derek de Solla Price. "The Water-Clock in the Tower of the Winds." *American Journal of Archaeology*, 72(1968):345–355, and figs. 111–118.

North, John D. "The Astrolabe." *Scientific American*, 230(January 1974):96–106.

North, John D. "Astrolabes and the Hour-line Ritual." *Journal for the History of Arabic Science*, 5(1981):113–114, repr. in his *Stars, Minds, and Fate: Essays in Ancient and Medieval Cosmology*. Ronceverte, W.V.: Hambledon Press, 1989.

North, John D. *Horoscopes and History*. Warburg Institute Surveys and Texts, 13. London: University of London, 1986.

North, John D. "Monasticism and the First Mechanical Clocks." In Julius T. Fraser and N. Lawrence, eds., *The Study of Time, II*. New York: Springer Verlag, 1975.

North, John D. "The Western Calendar – 'Intolerabilis, Horribilis, et Derisibilis': Four Centuries of Discontent." Pp. 75–113 in Coyne, Hoskin, and Pedersen, *Gregorian Reform of the Calendar*.

North, John D., and J. J. Roche, eds. *The Light of Nature: Essays in the History and Philosophy of Science Presented to A. C. Crombie*. Dordrecht: Martinus Nijhoff, 1985.

Ó Briain, Felim. "Brigitana." *Zeitschrift für Celtische Philologie*, 36(1977):112–137.

Ó Carragáin, Éamonn. "Crucifixion as Annunciation: The Relation of 'The Dream of the Rood' to the Liturgy Reconsidered." *English Studies*, 63(1982):487–505.

O'Connell, D. J. "Easter Cycles in the Early Irish Church." *Journal of the Royal Society of Antiquaries of Ireland*, 66(1936):67–106.

O'Connor, Elizabeth C. W. "The Star Mantle of Henry II." Ph.D. diss., Columbia University, 1980.

Ó Cróinín, Dáibhí. "The Irish Provenance of Bede's Computus." *Peritia*, 2(1983):229–247.

Ó Cróinín, Dáibhí. " 'New Heresy for Old': Pelagianism in Ireland and the Papal Letter of 640." *Speculum*, 60(1985):505–516.

O'Meara, John J. *Eriugena*. Oxford: Clarendon Press, 1988.

Obrist, Barbara. "Le diagramme isidorien des saisons, son contenu physique et les représentations figuratives." *Mélanges de l'École Française de Rome: Moyen âge*, 108/1(1996): 95–164.

Obrist, Barbara. "Wind Diagrams and Medieval Cosmology." *Speculum*, 72(1997):33–84.

Ogilvy, J. D. A. *Books Known to the English, 597–1066*. Cambridge: Mediaeval Academy of America, 1967.

Pedersen, Olaf. "The *Corpus Astronomicum* and the Traditions of Medieval Latin Astronomy: A Tentative Interpretation." Pp. 57–96 in [Owen Gingerich and Jerzy Dobrzycki, eds.,] *Colloquia Copernicana III*. Wroclaw: Ossolineum, 1975.

Pedersen, Olaf. "In Quest of Sacrobosco." *Journal for the History of Astronomy*, 16(1985): 175–221.

Pedersen, Olaf. "Some Astronomical Topics in Pliny." Pp. 162–196 in French and Greenaway, *Pliny, His Sources and Influence*.

Pedersen, Olaf. *A Survey of the Almagest*. Acta historica scientiarum naturalium et medicinalium, 30. Odense: Univ. Pr., 1974.

Pedersen, Olaf. "The *Theorica Planetarum* Literature of the Middle Ages." *Classica and Mediaevalia*, 23(1962):225–232.

Pingree, David. "Boethius' Geometry and Astronomy." Pp. 155–161 in Gibson, *Boethius*.

Pingree, David. "The Preceptum canonis Ptolomei." Pp. 355–375 in Jacqueline Hamesse and Marta Fattori, eds., *Rencontres de Cultures dans la Philosophie Médiévale: Traductions et Traducteurs de l'Antiquité Tardive au XIV^e Siècle*, Publications de l'Institut d'Études Médiévales–Textes, Études, Congrès, 11. Louvain-la-Neuve: Université Catholique de Louvain, 1990.

Poole, Rachael. "A Monastic Star Time Table of the Eleventh Century." *The Journal of Theological Studies*, 16(1915):98–104.

Poulle, Emmanuel. "Les instruments astronomiques de l'Occident Latin aux XI^e et XII^e siècles." *Cahiers de civilisation médiévale, X–XII^e siècles*, 15(1972):27–40.

Poulle, Emmanuel. "Le Traité de l'astrolabe d'Adélard de Bath." Pp. 199–132 in Burnett, *Adelard of Bath*.

Poulle, Emmanuel. "Walcher de Malvern et son astrolabe (1092)." *Rivista da Universidade de Coimbra*, 28(1980):47–54.

Prinz, Friedrich. *Frühes Mönchtum im Frankenreich: Kultur und Gesellschaft in Gallien, den Rheinland und Bayern am Beispiel der monastischen Entwicklung (4. bis 8. Jahrhundert)*. Munich and Vienna: R. Oldenbourg Verlag, 1965.

Rahner, Hugo, S. J. *Greek Myths and Christian Mystery*. Translated by Brian Battershaw. New York: Harper & Row, 1963.

Riché, Pierre. *Daily Life in the World of Charlemagne*. Translated by Jo Ann McNamara. Philadelphia: Univ. of Pennsylvania Pr., 1978.

Riché, Pierre. *Écoles et enseignement dans le Haut Moyen Age: Fin du V^e siècle – milieu du XI^e siècle*. 2^e ed. Paris: Picard, 1989.

Riché, Pierre. *Education and Culture in the Barbarian West: From the Sixth through the Eighth Century*. Translated by John Contreni. Columbia, S. C.: Univ. of South Carolina Pr., 1978.

Ridyard, Susan J. *The Royal Saints of Anglo-Saxon England: A Study of West Saxon and East Anglian Cults*. Cambridge: Cambridge Univ. Pr., 1988.

Ross, Anne. *Pagan Celtic Britain*. New York: Columbia Univ. Pr., 1967.

Ruggles, C. L. N., et al. *Megalithic Astronomy: A New Archaeological and Statistical Study of 300 Western Scottish Sites*. BAR British Series, 123. Oxford: B.A.R., 1984.

Ruggles, C. L. N., ed. *Records in Stone: Papers in Memory of Alexander Thom*. Cambridge: Cambridge Univ. Pr., 1988.

Ruggles, C. L. N., and N. J. Saunders, eds. *Astronomies and Cultures*. Niwot, Colo.: Univ. Pr. of Colorado, 1993.

Ruggles, C. L. N., and A. W. R. Whittle, eds. *Astronomy and Society in Britain during the Period 4000–1500 B.C.*, BAR British Series, 88. Oxford: B.A.R., 1981.

Samsó, Julio. "Sobre los materiales astronómicos en el 'Calendario de Córdoba' y en su version Latina del siglo XIII." Pp. 125–138 in Vernet, *Nuevos estudios sobre astronomía Española*.

Samsó, Julio. "La tradición clásica en los calendarios agrícolas Hispanoarabes y Norteafricanos." Pp. 177–186 in *Second International Congress of Studies on Cultures of the Western Mediterranean, Trabajos leidos en Barcelona, 29 Septembre–4 Octubre 1975*. Barcelona, 1978.

Sharpe, Richard. "*Vitae S. Brigitae*: The Oldest Texts." *Peritia*, 1(1982):81–106.

Silverstein, Theodore. "Daniel of Morley, English Cosmogonist and Student of Arabic Science." *Mediaeval Studies*, 10(1948):179–196.

Simon, Marcel. "Mithra, Rival du Christ?" Pp. 457–478 in *Études Mithraiques, Actes du 2ᵉ Congrès International, Téhéran*. Leiden: E. J. Brill, 1978.

Smalley, Beryl. *The Study of the Bible in the Middle Ages*. 2nd ed. Notre Dame, Ind.: Univ. of Notre Dame Pr., 1964.

Somerville, B. "Prehistoric Monuments in the Outer Hebrides and Their Astronomical Significance." *Journal of the Royal Anthropological Institute*, 42(1912):23–52.

Southern, R. W. *The Making of the Middle Ages*. New Haven: Yale Univ. Pr., 1953.

Southern, R. W. *Medieval Humanism*. New York: Harper Torchbooks, 1970.

Southern, R. W. *Robert Grosseteste: The Growth of an English Mind in Medieval Europe*. Oxford: Clarendon Press, 1986.

Stahl, William. *Martianus Capella and the Seven Liberal Arts*. 2 vols. Records of Civilization 84. New York: Columbia Univ. Pr., 1971–77.

Stahl, William. *Roman Science*. Madison: Univ. of Wisconsin Pr., 1962.

Stahl, William, ed. *Macrobius' Commentary on the Dream of Scipio*. Records of Civilization, 48. New York: Columbia Univ. Pr., 1952.

Stancliffe, Claire. *St. Martin and His Hagiographer: History and Miracle in Sulpicius Severus*. Oxford: Clarendon Press, 1983.

Stevens, Wesley M. "*Compotistica et astronomica* in the Fulda School." Pp. 27–67 in King and Stevens, *Saints, Scholars, and Heroes*, vol. 2.

Stevens, Wesley M. *Cycles of Time and Scientific Learning in Medieval Europe*. Variorum Collected Studies, CS 482. Aldershot: Variorum, 1995.

Stevens, Wesley M. "The Figure of the Earth in Isidore's 'De natura rerum.'" *Isis*, 71(1980):268–277.

Stiefel, Tina. "The Heresy of Science: A Twelfth-Century Conceptual Revolution." *Isis*, 68(1977):347–362.

Stock, Brian. *Myth and Science in the Twelfth Century: A Study of Bernard Silvester*. Princeton: Princeton Univ. Pr., 1972.

Strobel, August. *Texte zur Geschichte des frühchristlichen Osterkalendars*. Liturgiewissenschaftliche Quellen und Forschungen, 64. Münster i. W.: Verlag Aschendorff, 1984.

Strobel, August. *Ursprung und Geschichte des frühchristlichen Osterkalendars*, Texte und Untersuchungen zur Geschichte der altchristlichen Literatur, 121. Berlin: Akademie Verlag, 1977.

Stückelberger, Alfred. "Sterngloben und Sternkarten: Zur wissenschaftlichen Bedeutung des Leidener Aratus." *Museum Helveticum*, 47(1990):70–81.

Taft, Robert. *The Liturgy of the Hours in East and West*. Collegeville, Minn.: Liturgical Press, 1986.

Taub, Liba C. *Ptolemy's Universe: The Natural Philosophical and Ethical Foundations of Ptolemy's Astronomy*. Chicago: Open Court, 1993.

Tedlock, Barbara. *Time and the Highland Maya*, 2nd rev. ed. Albuquerque: Univ. of New Mexico Pr., 1992.

Tester, S. J. *A History of Western Astrology*. Woodbridge, U. K. and Wolfeboro, N. H.: Boydell Press, 1987.

Thom, Alexander. *Megalithic Lunar Observatories*. Oxford: Clarendon Press, 1971.

Thom, Alexander. *Megalithic Sites in Britain*. Oxford: Clarendon Press, 1967.

Thom, Alexander, and A. S. Thom. "A New Study of All Megalithic Lunar Lines." *Archaeoastronomy*, no. 2; suppl. to *Journal for the History of Astronomy*, 11(1980):S78– S89.

Thompson, J. W. "The Introduction of Arabic Science into Lorraine in the Tenth Century." *Isis*, 12(1929):184–193.

Thomson, Ron B. *Jordanus de Nemore and the Mathematics of Astrolabes:* De Plana Spera. Studies and Texts, 39. Toronto: Pontifical Institute of Mediaeval Studies, 1978.

Thorndike, Lynn. *A History of Magic and Experimental Science.* 8 vols. New York: Columbia Univ. Pr., 1923–58.

Thorndike, Lynn. *The* Sphere *of Sacrobosco and Its Commentators.* Chicago: Univ. of Chicago Pr., 1949.

Toomer, G. J. "The Solar Theory of az-Zarqāl: A History of Errors." *Centaurus*, 14(1969): 306–336.

Toomer, G. J. "The Solar Theory of az-Zarqāl: An Epilogue." Pp. 513–519 in King and Saliba, *From Deferent to Equant.*

Toomer, G. J. "A Survey of the Toledan Tables." *Osiris*, 15(1968):5–174.

Traube, Ludwig. "Computus Helperici." Pp. 128–156 in his *Vorlesungen und Abhandlungen,* vol. 3, Kleine Schriften. Munich: C. H. Beck, 1920.

Tuckerman, Bryant. *Planetary, Lunar, and Solar Positions . . . ,* Memoirs of the American Philosophical Society, vols. 56, 59. Philadelphia, 1962, 1964.

Turner, A. J. *The Time Museum: Catalogue of the Collection,* vol. 1, *Time Measuring Instruments,* part 1, *Astrolabes, Astrolabe Related Instruments.* Rockford, Ill.: The Time Museum, 1985.

Turton, D. A., and C. L. N. Ruggles. "Agreeing to Disagree: The Measurement of Duration in a Southwestern Ethiopian Community." *Current Anthropology*, 19(1978): 585–600.

Ulansey, David. "The Mythraic Mysteries." *Scientific American*, vol. 261, no. 6 (December 1989):130–135.

Ulansey, David. *The Origins of the Mithraic Mysteries: Cosmology and Salvation in the Ancient World.* Oxford: Oxford Univ. Pr., 1989.

Van Dam, Raymond. *Leadership and Community in Late Antique Gaul.* Berkeley and Los Angeles: Univ. of California Pr., 1985.

van de Vyver, A. "Les oeuvres inédites d'Abbon de Fleury." *Revue Bénédictine*, 43(1935): 125–169.

van de Vyver, A. "Les plus anciennes Traductions latines médiévales (Xᵉ–XIᵉ siècles) de Traités d'Astronomie et d'Astrologie." *Osiris*, 1(1936):658–691.

van Engen, John. "The Christian Middle Ages as an Historiographical Problem." *The American Historical Review*, 91(1986):519–552.

Vernet, Juan. "Les traductions scientifiques dans l'Espagne du Xᵉ siècle." *Cahiers de Tunisie*, 18(1970):47–59.

Vernet, Juan, ed. *Nuevos estudios sobre astronomía Española en el siglo de Alfonso X.* Barcelona: Instituto de Filologia, Institución 'Milá y Fontanals,' Consejo Superior de Investigaciones Cientificas, 1983.

Vernet, Juan, ed. *Textos y estudios sobre astronomía Española en el siglo XIII.* Barcelona: Facultad de Filosofia y Letras, Universidad Autónoma de Barcelona, 1981.

Vogel, Cyrille. *Introduction aux sources de l'histoire du culte Chrétien au moyen âge.* Biblioteca degli "Studi Medievali," 1. Spoleto, [1965].

Wallace-Hadrill, J. M. *Bede's Ecclesiastical History of the English People: A Historical Commentary.* Oxford: Clarendon Press, 1988.

Wallis, Faith. "MS Oxford, St. John's College 17: A Medieval Manuscript in Its Context." Ph.D. diss., University of Toronto, 1985.

Walsh, Maura, and Dáibhí Ó Cróinín, eds. *Cummian's Letter* De Controversia Paschali . . . *Together with a Related Irish Computistical Tract*, De Ratione Conputandi. Studies and Texts, 86. Toronto: Pontifical Institute of Mediaeval Studies, 1988.

Ward, Benedicta. *Miracles and the Medieval Mind: Theory Record and Event, 1000–1215*, rev. ed. Philadelphia: Univ. of Pennsylvania Pr., 1987.

Weisheipl, James A. "Curriculum of the Faculty of Arts at Oxford in the Early Fourteenth Century." *Mediaeval Studies*, 26(1964):143–185.

Welborn, Mary Catherine. "Lotharingia as a Center of Arabic and Scientific Influence in the Eleventh Century." *Isis*, 16(1931):188–199.

Whitelock, D., R. McKitterick, and D. Dumville. *Ireland in Early Medieval Europe: Studies in Memory of Kathleen Hughes*. Cambridge: Cambridge Univ. Pr., 1982.

Wiesenbach, Joachim. "Pacificus von Verona als Erfinder einer Sternenuhr." Pp. 229–250 in Butzer and Lohrmann, *Science in Carolingian Times*.

Wilson, Bryan R., ed. *Rationality*. New York: Harper & Row, 1970.

Zeilik, Michael. "The Ethnoastronomy of the Historic Pueblos, I: Calendrical Sun Watching." *Archaeoastronomy*, no. 8, suppl. to *Journal for the History of Astronomy*, 16(1985): S1–S24.

Zeilik, Michael. "The Ethnoastronomy of the Historic Pueblos, II: Moon Watching." *Archaeoastronomy*, no. 10, suppl. to *Journal for the History of Astronomy*, 17(1986):S1–S22.

Zinner, Ernst. "Die Tafeln von Toledo (Tabulae Toletanae)." *Osiris*, 1(1936):747–774.

# Index